CliffsNotes® Trigonometry Common Core Quick Review

CliffsNotes® Trigonometry Common Core Quick Review

By M. Sunil R. Koswatta, Ph.D.

About the Author

M. Sunil R. Koswatta, Ph.D., Mathematics, has been an educator for 25 years. Dr. Koswatta was a member of the EngageNY, New York State Education Department, Common Core curriculum writing team. He is currently a professor of Mathematics at Harper College in Illinois.

Acknowledgment

I would like to thank Richard Askey, University of Wisconsin, Madison; Hung-Hsi Wu, University of California, Berkeley; and Askey's MathEd listserv. I am grateful for the numerous discussions on K-12 mathematics education at MathEd.

Dedication

I would like to dedicate this book to Sunethra Koswatta, Erika Koswatta, and Swarna Kumudini Koswatta Atapattu.

Editorial

Executive Editor: Greg Tubach
Senior Editor: Christina Stambaugh
Senior Production Editor: Lisa Glover
Copy Editor: Lynn Northrup
Technical Editors: Mary Jane Sterling and Tom Page
Proofreader: Pamela Weber-Leaf
Indexer: Potomac Indexing, LLC

CliffsNotes® Common Core Trigonometry: Quick Review

Library of Congress Control Number: 2017950869

ISBN: 978-1-957671-16-1

Printed in the United States of America
DOC 10 9 8 7 6 5 4 3 2 1

Table of Contents

INTRODUCTION

The word *trigonometry* comes from Greek words meaning the measurement of triangles. Solving triangles is one of many aspects of trigonometry that you study today. To develop methods for solving triangles, trigonometric functions are constructed. The study of the properties of these functions and related applications forms the subject matter of trigonometry. Trigonometry has applications in navigation, surveying, construction, and many other branches of science, including mathematics and physics.

Trigonometry is an extension of algebraic thinking—particularly functions. The prerequisites for trigonometry comprehension include a familiarity with some topics from algebra and geometry. The algebra prerequisites include manipulating algebraic expressions, solving equation operations, determining the slope of a line, and solving and graphing linear equations. The geometry prerequisites include understanding properties of polygons, similar triangles, the Pythagorean Theorem, parallel and perpendicular lines, angle measurement, and volume and surface area.

If you feel that you need additional review of any of these topics, refer to *CliffsNotes Algebra II Common Core Quick Review* and *CliffsNotes Geometry Common Core Quick Review.*

Common Core State Standards for Mathematics (CCSSM)

CliffsNotes Trigonometry Common Core Quick Review topics are aligned to the Common Core State Standards for Mathematics (CCSSM)—the skills you should know and be able to perform in trigonometry. The concepts presented in this book are closely interrelated with a broader set of Common Core Mathematics conceptual categories:

- Numbers and quantity
- Algebra
- Functions
- Modeling

- Geometry
- Statistics and probability

The CCSSM define what you should know and what you should be able to execute using an integrated approach from each of these domains that builds on your strengths to critically think at a higher level of mathematics.

Connecting to Common Core Mathematics

CliffsNotes Trigonometry Common Core Quick Review gives you conceptual approaches to increase your knowledge, fluency, and skills in higher-level mathematics and real-world trigonometry applications. The standards of mathematical practices are observed in each chapter. In particular, the following Common Core State Standards math practices, MP3 and MP6, are relevant to trigonometry.

Construct viable arguments and critique the reasoning of others (MP3). Mathematically proficient students understand and use stated assumptions, definitions, and previously established results in constructing arguments. They make conjectures and build a logical progression of statements to explore the truth of their conjectures. They are able to analyze situations by breaking them into cases and can recognize and use counterexamples. They justify their conclusions, communicate them to others, and respond to the arguments of others. Students in all grades can listen to or read the arguments of others, decide whether they make sense, and ask useful questions to clarify or improve the arguments.

They reason inductively about data, making plausible arguments that take into account the context from which the data arose. Mathematically proficient students are also able to compare the effectiveness of two plausible arguments, distinguish correct logic or reasoning from that which is flawed, and—if there is a flaw in an argument—explain what it is. Elementary students can construct arguments using concrete referents such as objects, drawings, diagrams, and actions. Such arguments can make sense and be correct, even though they are not generalized or made formal until later grades. Later, students learn to determine domains to which an argument applies.

Attend to precision (MP6). Mathematically proficient students try to communicate precisely to others. They try to use clear definitions in discussion with others and in their own reasoning. They state the meaning of the symbols they choose, including using the equal sign, consistently and appropriately. They are careful about specifying units of measure and labeling axes to clarify the correspondence with quantities in a problem. They calculate accurately and efficiently and express numerical answers with a degree of precision appropriate for the problem context. In the elementary grades, students give carefully formulated explanations to each other. By the time they reach high school, they have learned to examine claims and make explicit use of definitions.

Why You Need This Book

Can you answer "yes" to any of these questions?

- Do you need to review the fundamentals of trigonometry fast?
- Do you need a course supplement to trigonometry?
- Do you need a concise, comprehensive reference for trigonometry?
- Do you need practice with real-life applications of trigonometry topics?

If so, then *CliffsNotes Trigonometry Common Core Quick Review* is for you!

How to Use This Book

Because mathematics builds on itself, many readers benefit most from studying or reviewing this book from cover to cover. However, you're the boss here, and you may choose to seek only the information you want and then put the book back on the shelf for later use. In that case, here are a few recommended ways to search for trigonometry topics:

- Look for your topic in the table of contents in the front of the book, or use the index to find specific topics.
- Flip through the book looking for subject areas by headings.
- Get a glimpse of what you'll gain from a chapter by reading through the "Chapter Check-In" and "Common Core Standard" references at the beginning of each chapter.
- Use the "Chapter Check-Out" at the end of each chapter to gauge your grasp of the important information you need to know.

- Test your knowledge more completely in the "Review Questions" (pp. 233–242).
- Look in the glossary (pp. 243–246) for important terms and definitions.
- Review the theorems in the appendix (pp. 247–258).

Trigonometry Theorems

There are numerous trigonometry theorems given in this book. They are numbered for organizational purposes. For example, *Theorem 2.2* is the second theorem given in Chapter 2. Study these theorems by their content, not by their number, as the theorem numbering has no significance outside of this book. As a handy study reference, the theorems are compiled in the appendix (pp. 247–258).

Hundreds of Practice Questions Online!

Go to CliffsNotes.com for hundreds of additional Trigonometry Common Core practice questions to help you prepare for your next quiz or test. The questions are organized by this book's chapter sections, so it is easy to use the book and then quiz yourself online to make sure you know the subject. Visit CliffsNotes.com to test yourself anytime and find other free homework help.

Chapter 1

GEOMETRIC PREREQUISITES

Chapter Check-In

- ❑ Recalling the definitions and properties of congruence, dilation, and similarity
- ❑ Recalling the congruence and similarity theorems of triangles
- ❑ Recalling the Pythagorean Theorem

Common Core Standard: Understanding similarity

Given two figures, use the definition of similarity in terms of similarity transformations to decide if they are similar; explain using similarity transformations the meaning of similarity for triangles as the equality of all corresponding pairs of angles and the proportionality of all corresponding pairs of sides. (HSG.SRT.A.2)

Use the properties of similarity transformations to establish the AA criterion for two triangles to be similar. (HSG.SRT.A.3)

Historically, trigonometry was developed to help find the measurements in triangles as an aid in navigation and surveying. Today, trigonometry is used in numerous sciences to help explain natural phenomena. In this chapter, we will look back at geometry prerequisites to study trigonometry. (See *CliffsNotes Geometry Common Core Quick Review* for detailed coverage of these topics.)

Trigonometry Theorems

The trigonometry theorems in this book are numbered for organizational purposes. For example, *Theorem 2.2* is the second theorem given in Chapter 2. Study these theorems by their content, not by their number, as the theorem numbering has no significance outside of this book.

Congruence

A rigid motion is a distance-preserving plane transformation. The basic rigid motions are translations, rotations, and reflections. A **congruence** in the plane is a plane transformation that is equal to the composition of a finite number of basic rigid motions. For two geometric figures S and R on the plane, we say S is congruent to R if there is a congruence C so that $C(S) = R$. We will denote this by the notation $S \cong R$. If $S \cong R$, then one can verify that $R \cong S$. That is, if there is a congruence that maps S to R, then there is another congruence that maps R to S.

The properties of the basic rigid motions are true for congruence.

Theorem 1.1: A congruence:
(i) maps lines to lines, rays to rays, and segments to segments.
(ii) maps line segments to line segments of equal length.
(iii) maps an angle to an angle of the same degree.

Theorem 1.2: A line segment is congruent to a line segment of equal length.

Theorem 1.3: An angle is congruent to an angle of the same degree.

Two triangles are congruent when the three sides and the three angles of one triangle have the same measurements as the three sides and the three angles of another triangle. Henceforth, we will refer to sides and angles of a triangle as *parts* of that triangle.

If two triangles are congruent, then the parts of the two triangles that have the same measurements are referred to as *corresponding parts.* This means that corresponding parts of congruent triangles are congruent by *Theorem 1.1*. Congruent triangles are named by listing their vertices in corresponding orders. For example, for the triangles below, we write $\Delta ABC \cong \Delta DEF$.

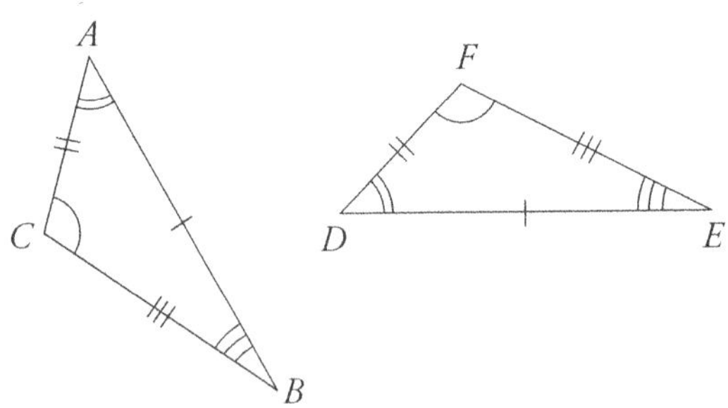

Congruent triangles

The knowledge of all parts of a triangle being congruent to corresponding parts of another triangle is not needed to prove that the two triangles are congruent to each other. It can be proven that if certain corresponding parts of triangles are congruent, then the other corresponding parts are also congruent. These results are known as congruence theorems.

*Theorem 1.4 (**SAS Theorem**):* Given two triangles ΔABC and ΔDEF so that $\angle A = \angle D$, $AB = DE$, and $AC = DF$, then the triangles are congruent.

*Theorem 1.5 (**ASA Theorem**):* Given two triangles ΔABC and ΔDEF so that $\angle A = \angle D$, $\angle B = \angle E$, and $AB = DE$, then the triangles are congruent.

*Theorem 1.6 (**SSS Theorem**):* Two triangles with three equal sides are congruent.

*Theorem 1.7 (**HL Theorem**):* If two right triangles have equal hypotenuses and one pair of equal legs, then the two triangles are congruent.

Dilation

Dilation is a plane transformation, but it is not a rigid motion. That is, dilations do not preserve distances. A dilation D with center O and scale factor r, where r is a positive number, maps each point P in the plane to a point $D(P)$ so that

(1) $D(O) = O$ and
(2) If $P \neq O$, then the point $D(P)$, denoted also by P', is the point on the ray $\overrightarrow{OP'}$ so that $OP' = r \cdot OP$.

(Note: A scale factor is a positive number in the definition of the dilation. See *CliffsNotes Geometry Common Core Quick Review* for detailed coverage of scale factors.)

The following figure shows the effect of a dilation with center O and scale factor 2 on a plane figure.

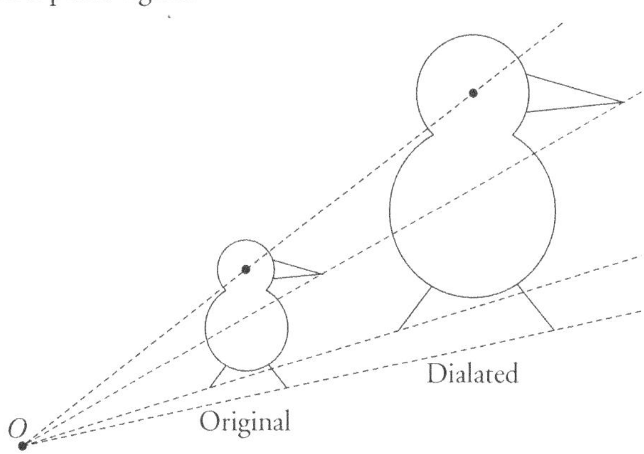

Theorem 1.8: A dilation:
(i) maps lines to lines, rays to rays, and segments to segments.
(ii) maps a line passing through the center O to itself.
(iii) maps a line not passing through O to a line parallel to it.

Theorem 1.9: A dilation preserves degrees of angles.

Similarity

A composition of a dilation and a congruence is called a **similarity.**

Let T and R be two figures in the plane. We say T is **similar** to R, denoted by $T \sim R$, if there is a dilation D and a congruence F so that $D(F(T)) = R$. The composition of a congruence F followed by a dilation D, that is, $D \circ F$, is called a similarity. Let S be the similarity $D \circ F$. That is, let $S = D \circ F$. If r is the scale factor of the dilation D, then we say r is also the scale factor of the similarity S.

Theorem 1.10: Any two circles are similar.

*Theorem 1.11 (**AA Criteria for Similarity**):* Two triangles with two pairs of equal angles are similar.

The converse of the AA Criteria for Similarity theorem is also true.

Theorem 1.12: If two triangles are similar, then the corresponding angles are congruent.

*Theorem 1.13 (**SSS Theorem for Similarity**):* If ΔABC and ΔPQR are triangles so that $\frac{AB}{PQ} = \frac{BC}{QR} = \frac{CA}{RP}$, then $\Delta ABC \sim \Delta PQR$.
The converse of the SSS Theorem for Similarity is also true.

Theorem 1.14: Suppose ΔABC and ΔPQR are triangles so that $\Delta ABC \sim \Delta PQR$. Then $\frac{AB}{PQ} = \frac{BC}{QR} = \frac{CA}{RP}$.

Theorem 1.15 (SAS Criteria for Similarity): Given two triangles ΔABC and ΔPQR, if $\angle A = \angle P$ and $\frac{AB}{PQ} = \frac{AC}{PR}$, then $\Delta ABC \sim \Delta PQR$.

Pythagorean Theorem

*Theorem 1.16 (**Pythagorean Theorem**):* Let ΔABC be a right triangle with $\angle B = 90°$. Then $AC^2 = AB^2 + BC^2$.

The proof of the Pythagorean Theorem using similarity is important enough that we will include it here.

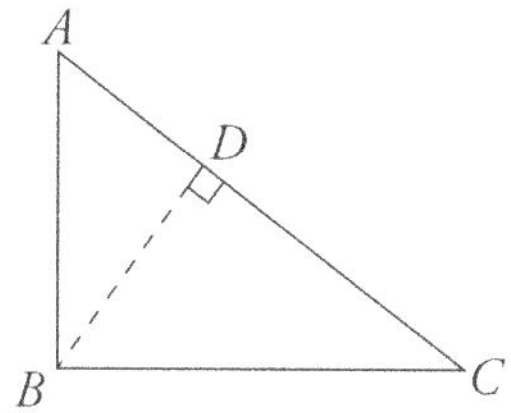

Drop a perpendicular from point B to line $\overleftrightarrow{AC}$. Let the foot of this perpendicular be point D. Then the triangles ΔABC and ΔADB are similar by the AA Criteria since the angle $\angle A$ is common to both right triangles. Then by *Theorem 1.14*, $\frac{AC}{AB} = \frac{AB}{AD}$. Using the cross-multiplication algorithm, we get $AB^2 = AC \cdot AD$.

The triangles ΔABC and ΔBDC are also similar by the AA Criteria since the angle $\angle C$ is common to both right triangles. By *Theorem 1.14*, $\frac{AC}{BC} = \frac{BC}{DC}$. Using the cross-multiplication algorithm, we get $BC^2 = AC \cdot DC$. Therefore, $AB^2 + BC^2 = AC \cdot AD + AC \cdot DC$.

Factoring out AC from the right side of the above equation, we get $AB^2 + BC^2 = AC(AD + DC)$. Since $AD + AC = AC$, we have the conclusion of the Pythagorean Theorem:

$$AB^2 + BC^2 = AC^2$$

The converse of the Pythagorean Theorem is also true.

Theorem 1.17 (***Converse of the Pythagorean Theorem***)*:* Suppose ΔABC is a right triangle so that $AB^2 + BC^2 = AC^2$. Then ΔABC is a right triangle with $\angle B = 90°$.

We will prove this theorem using the Pythagorean Theorem. Suppose $\angle B\ \pi\ 90°$. Drop a perpendicular from point A to line $\overleftrightarrow{BC}$ and let the foot of the perpendicular be point D.

There are two cases: Point D lies between points B and C or point B lies between points D and C. Proofs of both cases are similar. We will prove the case where point D lies between points B and C.

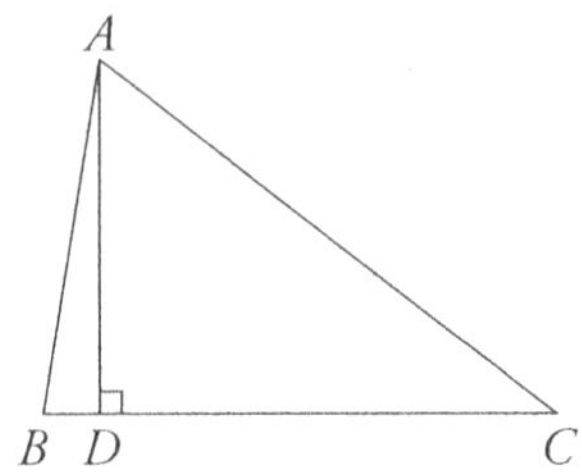

By the Pythagorean Theorem, $AD^2 + DC^2 = AC^2$. Since point D lies between points B and C, $BC > DC$ and therefore, $BC^2 > DC^2$.

By using the Pythagorean Theorem on the right triangle ΔABD, we get $AB^2 > AD^2$. Therefore, $AB^2 + BC^2 > AD^2 + DC^2 = AC^2$. However, this statement contradicts the hypothesis $AB^2 + BC^2 = AC^2$.

Therefore, our assumption that $\angle B \neq 90°$ is false; $\angle B = 90°$.

Chapter Check-Out

Questions

1. True or false: A congruence preserves the lengths of segments.
2. True or false: A similarity preserves the lengths of segments.
3. True or false: A congruence preserves the degrees of angles.
4. True or false: A similarity preserves the degrees of angles.

Answers

1. True
2. False
3. True
4. True

Chapter 2

TRIGONOMETRIC NUMBERS OF ANGLES

Chapter Check-In

- ❑ Understanding angles and signed measures
- ❑ Finding out about trigonometric numbers of an acute angle
- ❑ Defining trigonometric numbers of a general angle
- ❑ Identifying coordinates of a point on the unit circle by using trigonometric numbers of an angle
- ❑ Finding trigonometric numbers of special angles

Common Core Standard: Define trigonometric ratios

Understand radian measure of an angle as the length of the arc on the unit circle subtended by the angle. (HSF.TF.A.1)

Use special triangles to determine geometrically the values of sine, cosine, and tangent for $\frac{\pi}{3}$, $\frac{\pi}{4}$, and $\frac{\pi}{6}$ and use the unit circle to express the values of sine, cosine, and tangent for x, $\pi + x$, and $2\pi - x$ in terms of their values for x, where x is any real number. (HSF.TF.A.3)

Understand that by similarity, side ratios in right triangles are properties of the angles in the triangle, leading to definitions of trigonometric ratios for acute angles. (HSG.SRT.C.6)

Explain and use the relationship between the sine and cosine of complementary angles. (HSG.SRT.C.7)

Please also refer to the standards for mathematical practice, found here: www.corestandards.org/Math/Practice/.

In this chapter, we will define angle measure and basic trigonometric relationships. The whole discussion is restricted to a plane (2-space).

Trigonometry Theorems

The trigonometry theorems in this book are numbered for organizational purposes. For example, *Theorem 2.2* is the second theorem given in Chapter 2. Study these theorems by their content, not by their number, as the theorem numbering has no significance outside of this book.

Angles

Introduce a Cartesian coordinate system to 2-space so that one of the axes is known as the x-axis and the other is known as the y-axis. We will follow the usual convention of choosing the x-axis to be horizontal and the y-axis to be vertical. Now, consider an angle in the plane. Use a rigid motion to move the angle so that one of the sides of the angle is on the positive x-axis and the vertex is at the origin. We say that such an angle is in **standard position.**

Now, we will think of this **angle** as a rotation around the origin by identifying the side on the positive x-axis as the **initial side** and the other side as the **terminal side.** That is, the rotation is starting from the initial side and ending at the terminal side. We will assign a signed measure to the rotation as follows: If the direction of the rotation is counterclockwise, then the measure of the rotation is the measure of the angle with a positive sign, a **positive angle.** If the direction of the rotation is clockwise, then the measure of the rotation is the measure of the angle with a negative sign, a **negative angle.**

Angles can be measured in one of two units: degrees or radians. The relationship between these two measures may be expressed as follows:

$$180^\circ = \pi \text{ radians}$$

$$1^\circ = \frac{\pi}{180} \text{ radians}$$

$$\text{or,} \quad 1 \text{ radian} = \frac{180^\circ}{\pi}$$

The fact that 180° is the same as π radians is extremely important. From this relationship, the following proportion can be used to convert between **radian measure of an angle** and **degree measure of an angle:**

$$\frac{\theta^\circ}{180^\circ} = \frac{\theta \text{ rad}}{\pi \text{ rad}}$$

Example 1: What is the degree measure of a 2.4 radian angle?

$$\frac{\theta}{180^\circ} = \frac{2.4}{\pi}$$
$$\theta = \frac{(180^\circ)(2.4)}{\pi} = \frac{432^\circ}{\pi}$$
$$\theta \approx 137.5^\circ$$

Example 2: What is the radian measure of a 63° angle?

$$\frac{63^\circ}{180^\circ} = \frac{\theta}{\pi}$$
$$\theta = \frac{(\pi)(63^\circ)}{180^\circ} = \frac{7\pi}{20} \text{ rad}$$
$$\theta \approx 1.1 \text{ rad}$$

The radian measures of certain special angles follow directly from the radian-degree relationships. Some of these are summarized in Table 2-1.

Table 2-1 Degree/Radian Equivalencies

Degrees	*0°*	*30°*	*45°*	*60°*	*90°*	*120°*	*135°*	*150°*	*180°*
radians	0	$\frac{\pi}{6}$	$\frac{\pi}{4}$	$\frac{\pi}{3}$	$\frac{\pi}{2}$	$\frac{2\pi}{3}$	$\frac{3\pi}{4}$	$\frac{5\pi}{6}$	π

As a convention, we will use lowercase Greek letters such as α, β, γ, and θ to identify angles (rotations). We will also use the signed measure of the angle to identify angles (rotations). The following figure shows two examples of angles.

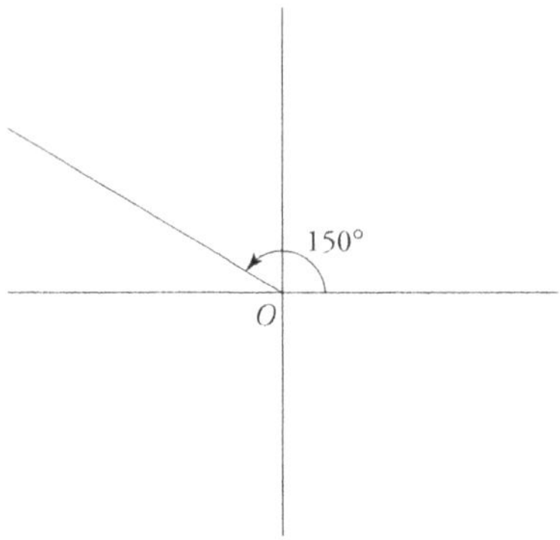

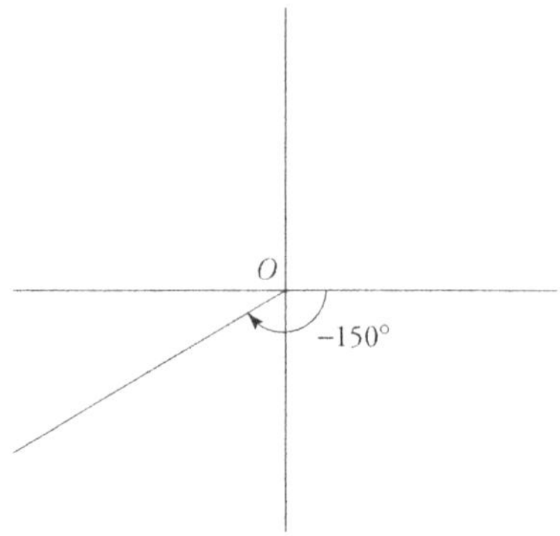

An angle that is in standard position is said to be a **quadrantal angle** if its terminal side coincides with a coordinate axis. For example, one complete counterclockwise rotation is measured as 360° (or 2π radians). (In this case, the terminal side coincides with the initial side.) We will identify an angle in standard position by the quadrant where the terminal side is located. For example, if the terminal side is in the third quadrant, then we say the angle is in the third quadrant. The following figures show some examples.

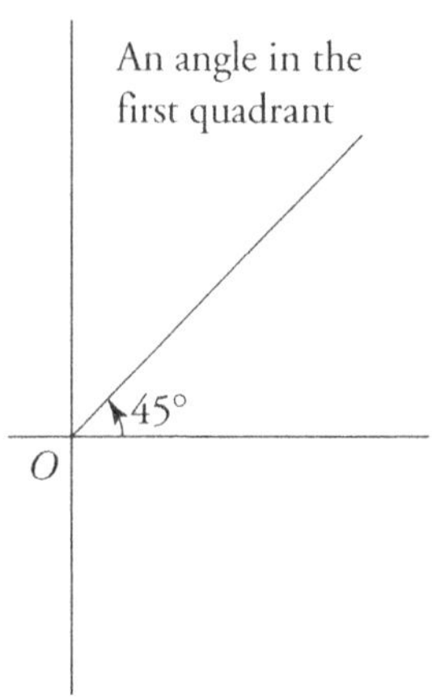

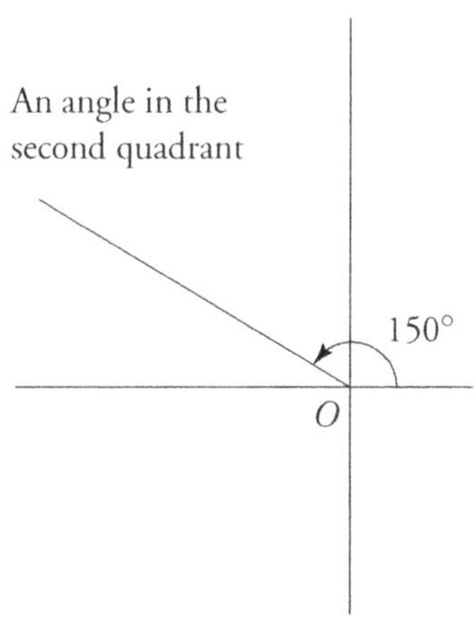

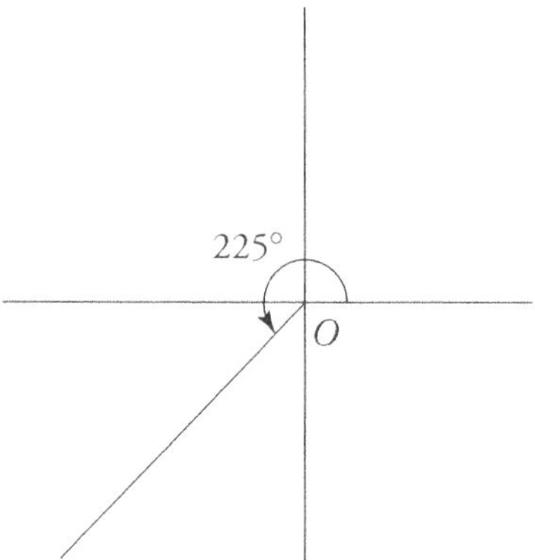

An angle in the third quadrant

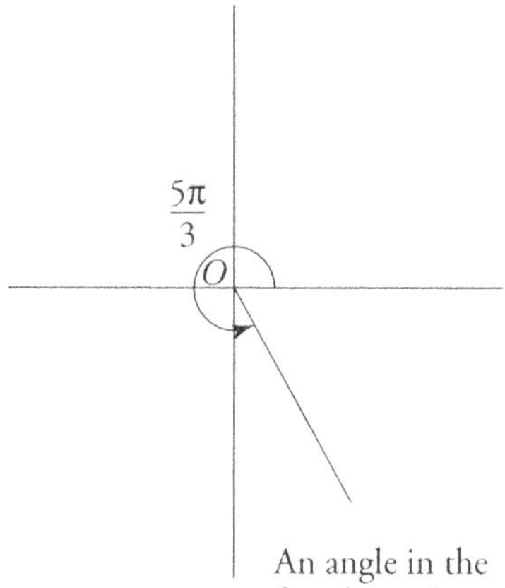

An angle in the fourth quadrant

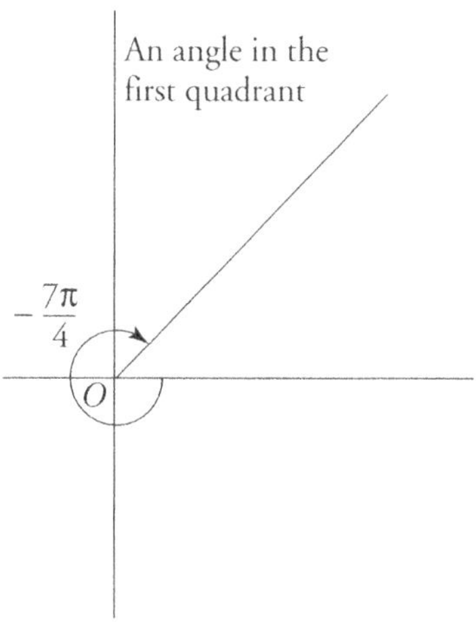

An angle in the first quadrant

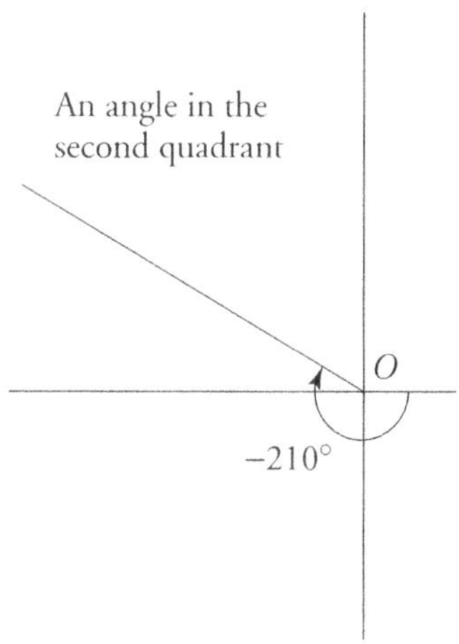

An angle in the second quadrant

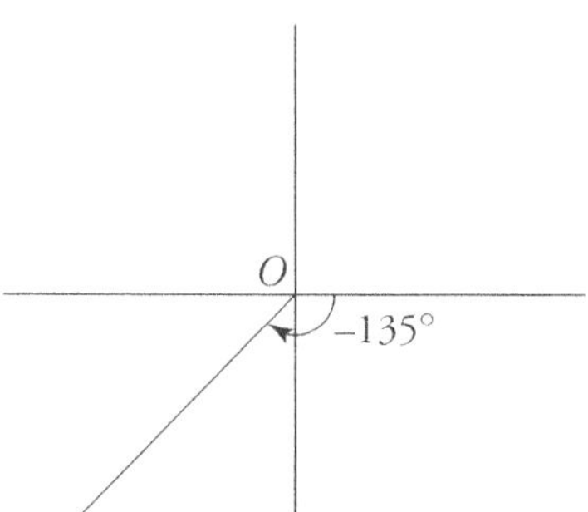

An angle in the third quadrant

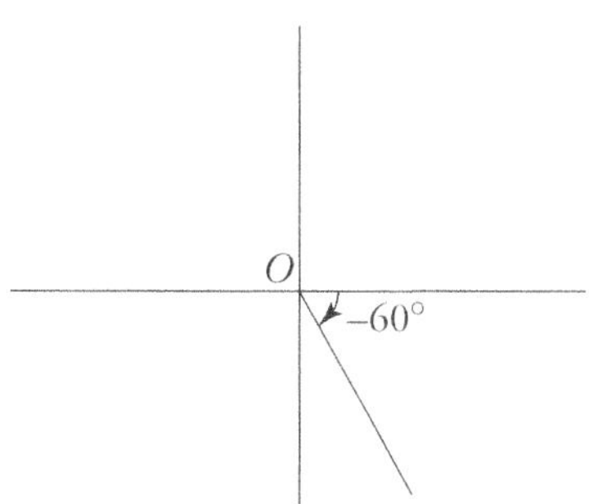

An angle in the fourth quadrant

Example 3: The following angles (in standard position) terminate in the listed quadrant.

$94°$	second quadrant
$\frac{25\pi}{9}$	second quadrant
$-100°$	third quadrant
π	quadrantal
$-300°$	first quadrant

Two angles in standard position that share a common terminal side are said to be **coterminal.** Clearly, the angle 0 (in radians) is coterminal with any of the angles 2π, 4π, 6π, etc. and -2π, -4π, -6π, etc. The following theorem follows immediately.

Theorem 2.1: Let θ be an angle in standard position. Then θ and $\theta + 2n\pi$ are coterminal angles for any integer n.

Example 4: Is an angle measuring 200° coterminal with an angle measuring 940°?

If an angle measuring 940° and an angle measuring 200° are coterminal, then there is an integer n so that

$$940 = 200 + 360n$$

This linear equation in n can be solved for n as shown below.

$$740 = 360n$$
$$n = \frac{740}{360}$$
$$n = \frac{37}{18}$$

This shows that there is no such integer as n. Therefore, the two angles are not coterminal.

The angles in the following figures are all coterminal with 30°.

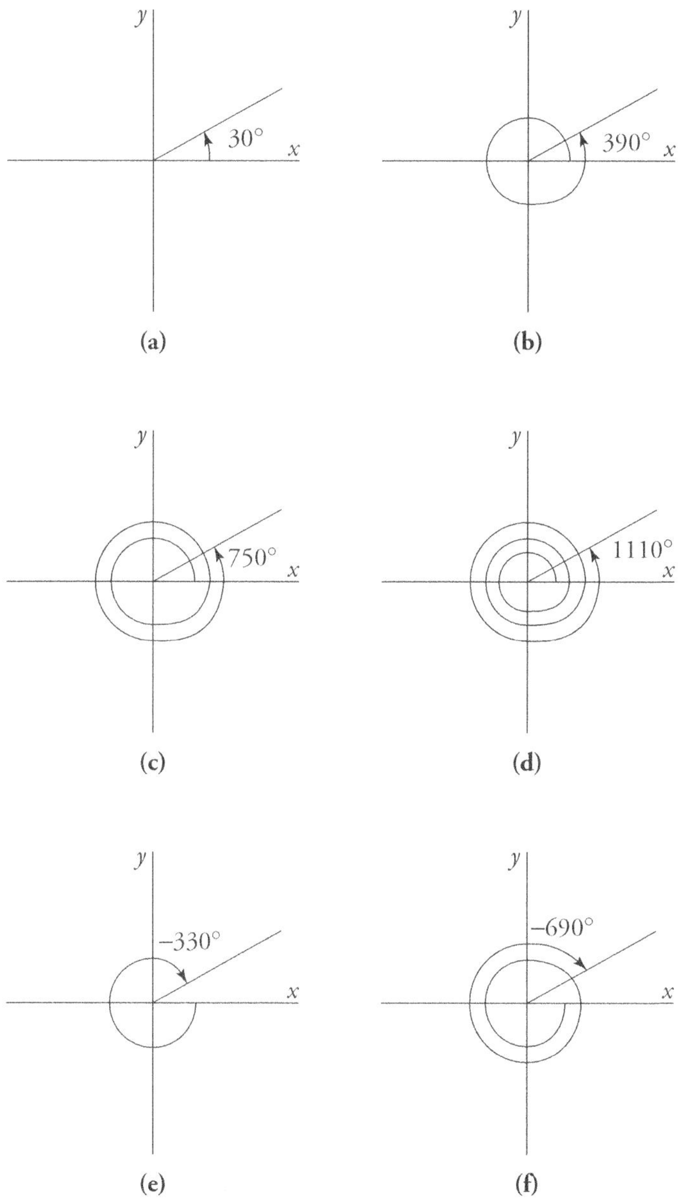

Example 5: Name five angles that are coterminal with –70°.

$$-70° + (1)360° = 290°$$
$$-70° + (2)360° = 650°$$
$$-70° + (3)360° = 1{,}010°$$
$$-70° + (-1)360° = -430°$$
$$-70° + (-2)360° = -790°$$

Sine and Cosine Numbers of Acute Angles

Consider an angle θ in the standard position with a positive measure less than $\frac{\pi}{2}$. Pick an arbitrary point P(x, y) on the terminal side other than O. Draw a perpendicular from P to the positive x-axis, and let the foot of the perpendicular be Q as shown in the following figure. Then ΔOPQ is a right triangle, where θ is an acute angle. The opposite side of θ has length y, and the adjacent side of θ has length x. By the distance formula for 2-space, the hypotenuse has the length $\sqrt{x^2 + y^2}$. We will use the letter r to represent the length of the hypotenuse for convenience. That is, $r = \sqrt{x^2 + y^2}$.

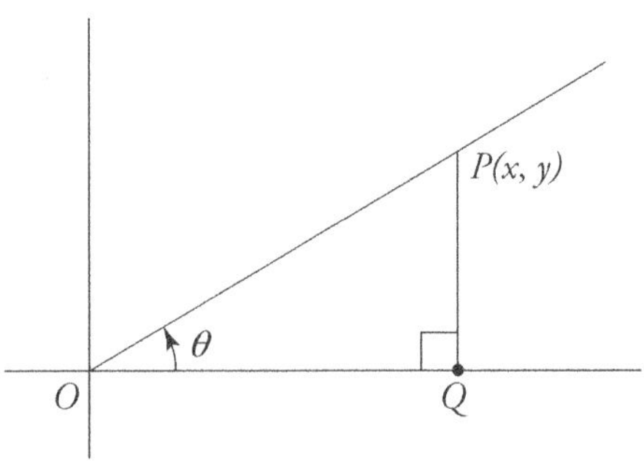

We define two numbers (ratios) associated with θ, namely sine θ (written sin θ) and cosine θ (written cos θ) as follows:

$$\sin\theta = \frac{y}{r} \text{ and } \cos\theta = \frac{x}{r}$$

With reference to the right triangle ΔOPQ, these definitions can be viewed as $\sin\theta = \dfrac{\text{length of the side opposite of } \theta}{\text{length of the hypotenuse}}$ and

$$\cos\theta = \frac{\text{length of the side adjacent to } \theta}{\text{length of the hypotenuse}}.$$

It seems that both definitions of sin θ and cos θ depend on the chosen point P on the terminal side of θ. However, the following theorem shows that these numbers do not depend on the chosen point. In that sense, both sin θ and cos θ are unique for a given θ.

Theorem 2.2: Let θ be an angle in standard position with a positive measure less than $\dfrac{\pi}{2}$. Then both sin θ and cos θ are unique.

Proof: We will prove the uniqueness of sin θ. The proof of the uniqueness of cos θ is similar. Let $P_1(x_1, y_1)$ and $P_2(x_2, y_2)$ be two distinct arbitrary points on the terminal side of θ other than O. Then, by the definition of sin θ, $\sin\theta = \dfrac{y_1}{r_1}$, where $r_1 = \sqrt{x_1^{\ 2} + y_1^{\ 2}}$, and $\sin\theta = \dfrac{y_2}{r_2}$, where $r_2 = \sqrt{x_2^{\ 2} + y_2^{\ 2}}$.

See the following figure. If sin θ is unique, then $\dfrac{y_1}{r_1}$ must equal $\dfrac{y_2}{r_2}$. If we can show that, then the arbitrary choice of the points indicates that sin θ is the same for any chosen point on the terminal side of θ other than O.

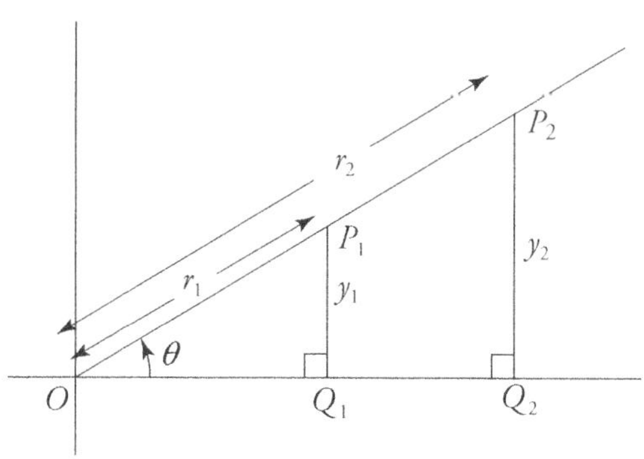

Draw a perpendicular from P_1 to the positive x-axis. Let the foot of this perpendicular be Q_1. Draw a perpendicular from P_2 to the positive x-axis. Let the foot of this perpendicular be Q_2. The triangles ΔOP_1Q_1 and ΔOP_2Q_2 are similar by the AA Criteria for Similarity (*Theorem 1.11*). Then we get

$$\frac{P_1Q_1}{OP_1} = \frac{P_2Q_2}{OP_2}$$

$$\frac{y_1}{r_1} = \frac{y_2}{r_2}$$

Therefore, sin θ is unique for a given positive θ less than $\frac{\pi}{2}$.

Sine and Cosine Numbers of an Acute Angle of a Right Triangle

Consider any right triangle ΔPQR so that the angle $\angle PQR$ is the right angle. Let θ be one of the acute angles of the right triangle; say, $\angle PRQ$. Introduce a Cartesian coordinate system to the plane. Let C be a congruence such that $C(R) = O$, $C(Q) = Q_1$ so that Q_1 is a point on the positive x-axis, and $C(P) = P_1$. Now the adjacent side of the angle θ in the triangle ΔP_1Q_1O is on the positive x-axis.

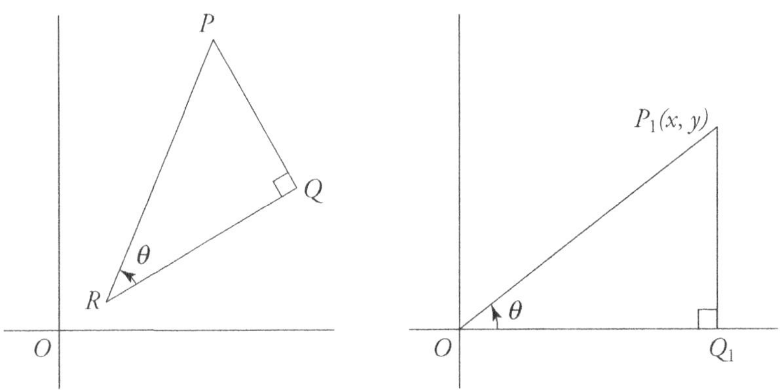

Suppose the coordinates of P_1 are (x, y). Then $x = RQ$ and $y = PQ$. Let $PR = r$. Then by the definitions of sin θ and cos θ, $\sin\theta = \frac{y}{r} = \frac{P_1Q_1}{r}$ and $\cos\theta = \frac{x}{r} = \frac{OQ_1}{r}$. Therefore, we have the following theorem.

Theorem 2.3: Suppose θ is an acute angle of a right triangle. Then

$$\sin\theta = \frac{\text{length of the opposite side of } \theta}{\text{length of the hypotenuse}}$$

and

$$\cos\theta = \frac{\text{length of the adjacent side of } \theta}{\text{length of the hypotenuse}}.$$

Let θ and α be the acute angles of a right triangle. Then $\alpha = 90° - \theta$ and α is the complement of θ. By *Theorem 2.3*, sin θ = cos $(90° - \theta)$ and cos θ = sin $(90° - \theta)$.

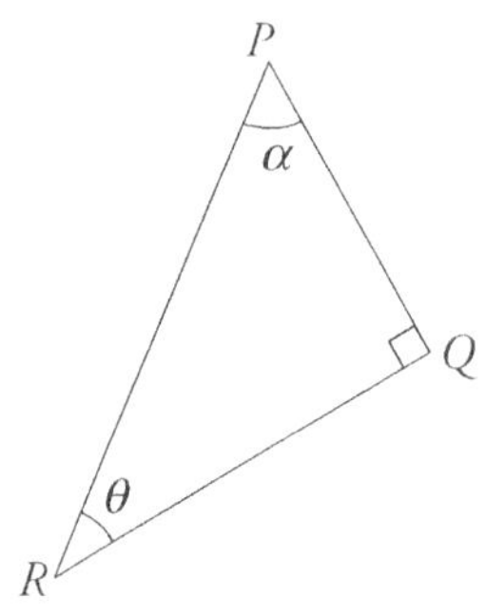

Example 6: Consider the right triangle ΔPQR where $\angle PQR$ is the right angle. Suppose $\angle PRQ = \alpha$, $PQ = 3$, and $QR = 4$. Find sin α and cos α.

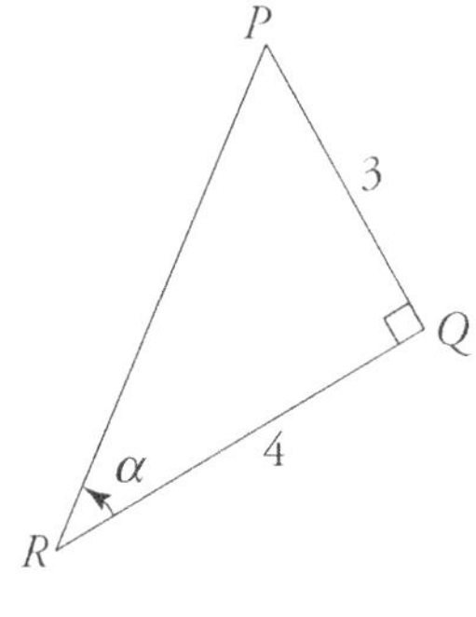

By the Pythagorean Theorem,

$$PR^2 = PQ^2 + QR^2$$
$$PR^2 = 3^2 + 4^2$$
$$PR^2 = 9 + 16$$
$$PR^2 = 25$$

Since PR is a length, $PR = 5$.

By *Theorem 2.3*, $\sin \alpha = \frac{3}{5}$ and $\cos \alpha = \frac{4}{5}$.

Example 7: Consider the right triangle ΔPQR, where $\angle PQR$ is the right angle. Suppose $\angle PRQ = \alpha$, $\angle QPR = \beta$, $PQ = \frac{3}{5}$, and $QR = \frac{4}{5}$. Find $\sin \alpha$, $\cos \alpha$, $\sin \beta$, and $\cos \beta$.

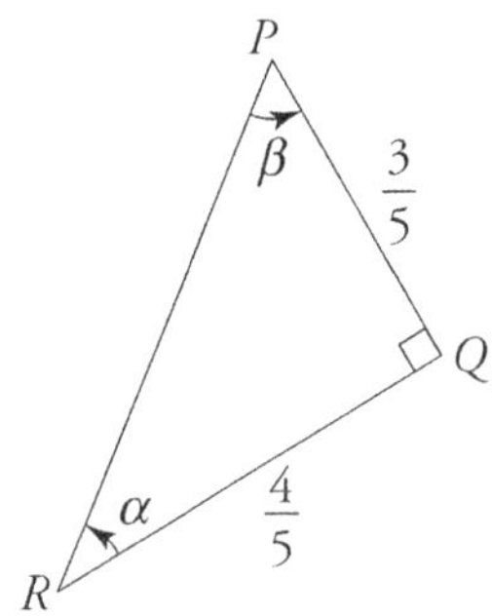

By the Pythagorean Theorem,

$$PR^2 = PQ^2 + QR^2$$
$$PR^2 = \left(\frac{3}{5}\right)^2 + \left(\frac{4}{5}\right)^2$$
$$PR^2 = \frac{9}{25} + \frac{16}{25}$$
$$PR^2 = 1$$

Since PR is a length, $PR = 1$.

Then by *Theorem 2.3*, $\sin\alpha = \frac{3}{5}$, $\cos\alpha = \frac{4}{5}$, $\sin\beta = \frac{4}{5}$, and $\cos\beta = \frac{3}{5}$.

Sine and Cosine Numbers of Any Angles

Let θ be any angle in standard position. Let $P(x, y)$ be any arbitrary point on the terminal side of θ other than O. Let $OP = r$. We define $\sin\theta$ and $\cos\theta$ as follows:

$$\sin\theta = \frac{y}{r} \text{ and } \cos\theta = \frac{x}{r}$$

Notice that $r = \sqrt{x^2 + y^2}$ by the two-dimensional distance formula. (Note: From now on, we will refer to the two-dimensional distance formula as simply "distance formula.")

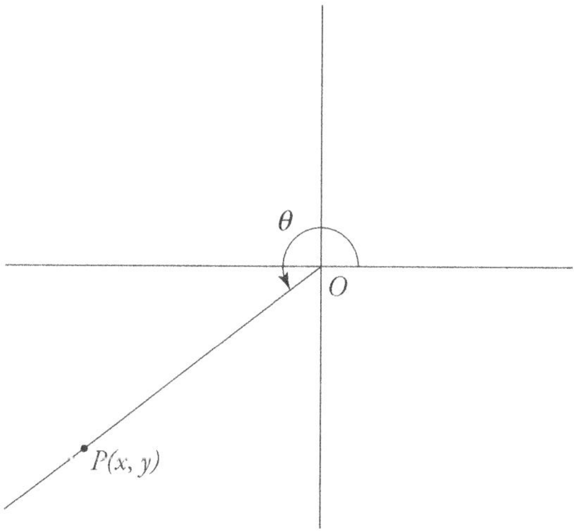

Theorem 2.4: For any angle θ in standard position, the numbers $\sin\theta$ and $\cos\theta$, as defined above, are unique.

Even though we use the following figure to facilitate the proof of *Theorem 2.4*, the proof does not depend on the figure or the angle in the figure. Also, we will prove that $\cos\theta$ is unique, and the proof of $\sin\theta$ is unique and similar.

Proof: Let $P_1(x_1, y_1)$ and $P_2(x_2, y_2)$ be any two distinct points on the terminal side of θ other than O. Draw perpendiculars from P_1 and P_2 to the x-axis and let the feet of these perpendiculars be Q_1 and Q_2, respectively.

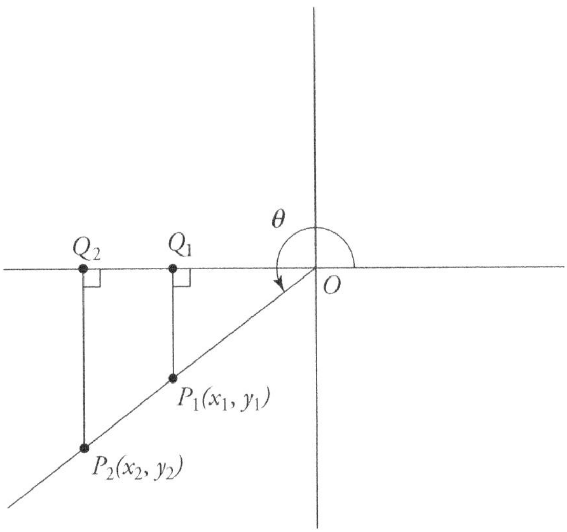

Angles $\angle P_1OQ_1$ and $\angle P_2OQ_2$ are equal (common angles), and angles $\angle P_1Q_1O$ and $\angle P_2Q_2O$ are equal (right angles). Therefore, by the AA Criteria for Similarity (*Theorem 1.11*), triangles ΔOP_1Q_1 and ΔOP_2Q_2 are similar. Therefore,

$$\frac{OQ_1}{OP_1} = \frac{OQ_2}{OP_2}$$

$$\frac{|x_1|}{r_1} = \frac{|x_2|}{r_2}$$

Since P_1 and P_2 lie in the same quadrant, the signs of x_1 and x_2 are the same. Therefore, $\frac{x_1}{r} = \frac{x_2}{r}$.

Since P_1 and P_2 are arbitrary points, this proves that $\cos\theta$ is unique.

Theorem 2.5: Suppose θ is an angle in standard position. Then the following are true:
(1) The angle θ is in the first quadrant if and only if $\sin\theta > 0$ and $\cos\theta > 0$.
(2) The angle θ is in the second quadrant if and only if $\sin\theta > 0$ and $\cos\theta < 0$.
(3) The angle θ is in the third quadrant if and only if $\sin\theta < 0$ and $\cos\theta < 0$.
(4) The angle θ is in the fourth quadrant if and only if $\sin\theta < 0$ and $\cos\theta > 0$.

Proof: Let $P(x, y)$ be an arbitrary point on the terminal side of θ other than O and let $r = OP$. Suppose also that θ is not quadrantal. By the distance formula, $r = \sqrt{x^2 + y^2} > 0$. If θ is in the first quadrant, then both x and y are positive and vice versa. Therefore, by definition, both $\sin\theta$ and $\cos\theta$ are positive. If θ is in the second quadrant, then x is negative and y is positive and vice versa. Therefore, by definition, $\sin\theta$ is positive and $\cos\theta$ is negative. If θ is in the third quadrant, then both x and y are negative and vice versa. Therefore, by definition, both $\sin\theta$ and $\cos\theta$ are negative. If θ is in the fourth quadrant, then x is positive and y is negative and vice versa. Therefore, by definition, $\sin\theta$ is negative and $\cos\theta$ is positive.

Example 8: Suppose θ is an angle in standard position and θ is in the second quadrant. Suppose $P(-2, 3)$ is a point on the terminal side of θ. Find $\sin\theta$ and $\cos\theta$ exactly.

Let $r = OP$. Then by the distance formula,

$$r = \sqrt{(-2-0)^2 + (3-0)^2}$$
$$r = \sqrt{(-2)^2 + 3^2}$$
$$r = \sqrt{4+9} = \sqrt{13}$$

By definition, $\sin\theta = \dfrac{3}{\sqrt{13}}$ and $\cos\theta = \dfrac{-2}{\sqrt{13}}$.

The Unit Circle

The circle with center O and radius 1 is called the **unit circle.** Given an angle θ in standard position, we can choose any point (other than O) on the terminal side of θ to define $\cos\theta$ and $\sin\theta$. We will choose the point of intersection of the terminal side of θ and the unit circle. Let $P(x, y)$ be that point. Then $OP = 1$. Then by definition, $\sin \theta = y$ and $\cos \theta = x$. In other words, any point P on the unit circle has coordinates $(\cos \theta, \sin \theta)$ for some angle θ.

Theorem 2.6: Any point P on the unit circle has coordinates $(\cos \theta, \sin \theta)$ for some angle θ in standard position, where P is the point of intersection of the unit circle and the terminal side of θ.

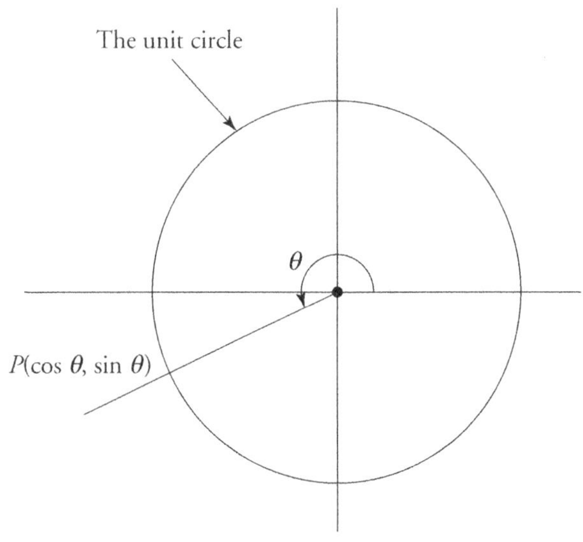

It is a convention to write $\sin^2 \theta$ for $(\sin \theta)^2$.

Theorem 2.7: If α and β are coterminal angles, then $\cos \alpha = \cos \beta$ and $\sin \alpha = \sin \beta$.

This follows directly from *Theorem 2.6.*

Theorem 2.8: For any angle θ, $\sin^2 \theta + \cos^2 \theta = 1$.

Proof: Let θ be any given angle. Use a rigid motion, if necessary, and put θ in standard position. Then $P(\cos \theta, \sin \theta)$ is a point on the unit circle. By the distance formula,

$$OP = \sqrt{(\cos\theta - 0)^2 + (\sin\theta - 0)^2}$$

$$1 = \sqrt{\cos^2 \theta + \sin^2 \theta}$$

By squaring both sides of the equation above, we get $\sin^2 \theta + \cos^2 \theta = 1$.

Theorem 2.9: For any angle θ and for any integer n:
(1) $\sin(2n\pi + \theta) = \sin \theta$ and
(2) $\cos(2n\pi + \theta) = \cos \theta$.

Proof: Without loss of generality, assume both angles θ and $2n\pi + \theta$ are in standard position for a given integer n. Then θ and $2n\pi + \theta$ are coterminal angles by *Theorem 2.1*. Then $P(\cos \theta, \sin \theta)$ and $Q(\cos(2n\pi + \theta), \sin(2n\pi + \theta))$ are the same point on the unit circle. Therefore, $\cos(2n\pi + \theta) = \cos \theta$ and $\sin(2n\pi + \theta) = \sin \theta$.

Theorem 2.10: For any angle θ:
(1) $\sin(-\theta) = -\sin \theta$ and
(2) $\cos(-\theta) = \cos \theta$.

Proof: Let $P(\cos \theta, \sin \theta)$ and $Q(\cos(-\theta), \sin(-\theta))$ be points on the unit circle.

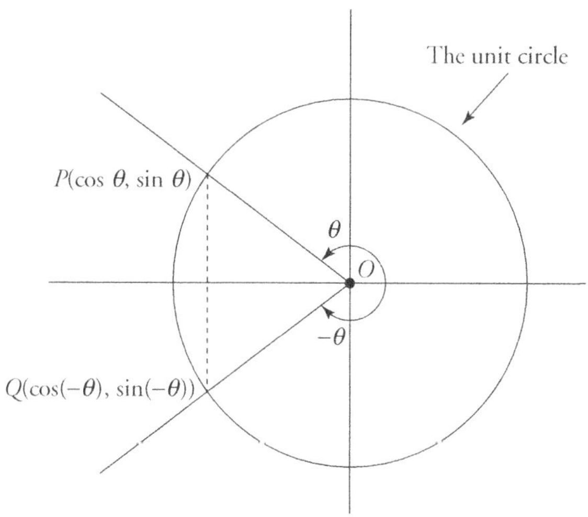

Let R be the reflection across the x-axis. Since reflections preserve the measures of angles (initial sides of both θ and $-\theta$ are the same, and the measures $|-\theta| = \theta$ are the same), the reflected ray $\overrightarrow{OP}$ is the ray $\overrightarrow{OQ}$; that is, $R(\overrightarrow{OP}) = \overrightarrow{OQ}$. Reflections also preserve lengths, and since $OP = OQ$ and $R(O) = O$, the reflected image of P is Q. Therefore, the line $\overleftrightarrow{PQ}$ is perpendicular to the x-axis. Since ΔPOQ is an isosceles triangle, the x-axis is the perpendicular bisector of the segment $\overline{PQ}$. That is, the y-coordinate of the point P is the negative y-coordinate of the point Q. That is, $-\sin(-\theta) = \sin\theta$ or $\sin(-\theta) = -\sin\theta$.

Also, the x-coordinates of P and Q are the same. That is, $\cos(-\theta) = \cos\theta$.

Theorem 2.11: For any angle θ:
(1) $\sin(\pi - \theta) = \sin\theta$ and
(2) $\cos(\pi - \theta) = -\cos\theta$.

Proof:

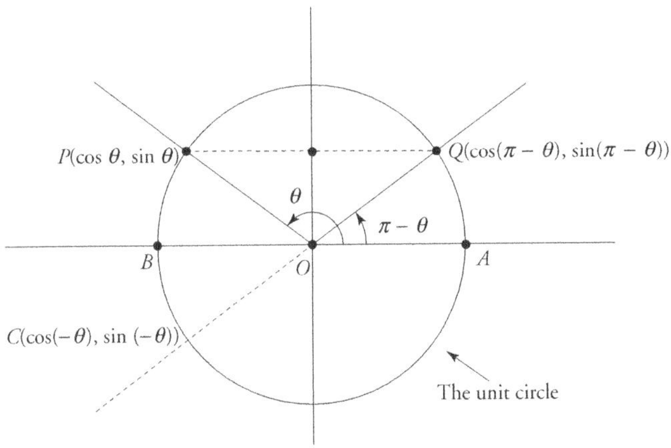

Suppose θ is in standard position. To make the argument easy to follow, we identify the following points: Let A be the point of intersection of the unit circle and the positive x-axis, let B be the point of intersection of the unit circle and the negative y-axis, let C be the point with coordinates $(\cos(-\theta), \sin(-\theta))$, and let Q be the point with coordinates $(\cos(\pi - \theta), \sin(\pi - \theta))$. Then points C and Q lie on the same line λ, since $m\angle COQ = \pi$ and the points lie on the opposite rays separated by O. Since $m\angle AOQ = (\pi - \theta)$, the clockwise angle $\angle BOQ$ is $-\theta$.

Let R be the reflection across the y-axis. Since reflections preserve degrees of angles, ray $\overrightarrow{OP}$ is reflected onto ray $\overrightarrow{OQ}$, and since reflections preserve lengths of segments, point P is reflected onto point Q. Therefore, line $\overleftrightarrow{PQ}$ is perpendicular to the y-axis. Since ΔPOQ is an isosceles triangle, the y-axis is the perpendicular bisector of segment $\overline{PQ}$. That is, the x-coordinate of point P is the negative x-coordinate of point Q. That is, $\cos\theta = -\cos(\pi - \theta)$, or $\cos(\pi - \theta) = -\cos\theta$.

Also, the y-coordinates of points P and Q are the same. That is, $\sin(\pi - \theta) = \sin\theta$.

Example 9: Find sin θ if θ is an acute angle ($0° < \theta < 90°$) and $\cos\theta = \frac{1}{4}$.

By *Theorem 2.8*, $\sin^2\theta + \cos^2\theta = 1$. Therefore,

$$\sin^2\theta + \left(\frac{1}{4}\right)^2 = 1$$

$$\sin^2\theta = 1 - \frac{1}{16}$$

$$\sin^2\theta = \frac{15}{16}$$

$$\sin\theta = \pm\frac{\sqrt{15}}{4}$$

Since θ is in the first quadrant, $\sin\theta = \frac{\sqrt{15}}{4}$ by *Theorem 2.5*.

Example 10: Suppose $\sin\theta < 0$ and $\cos\theta = \frac{1}{4}$. Find $\sin\theta$.

Since $\sin\theta < 0$ and $\cos\theta > 0$, the angle θ is in the fourth quadrant by *Theorem 2.5*. Then by *Theorem 2.8*,

$$\sin^2\theta + \cos^2\theta = 1$$

$$\sin^2\theta + \left(\frac{1}{4}\right)^2 = 1$$

$$\sin^2\theta = 1 - \frac{1}{16}$$

$$\sin^2\theta = \frac{15}{16}$$

$$\sin\theta = \pm\frac{\sqrt{15}}{4}$$

Since θ is in the fourth quadrant, $\sin\theta = -\frac{\sqrt{15}}{4}$.

Sine and Cosine Numbers of Special Angles

The angles whose radian measures are 0, $\frac{\pi}{6}$, $\frac{\pi}{4}$, $\frac{\pi}{3}$, and $\frac{\pi}{2}$ are called **special angles.** These angles are very important in the study of trigonometry; it is wise to remember their measures in both radians and degrees. The special angles in degrees are 0°, 30°, 45°, 60°, and 90° (see Table 2-1 on p. 13). We will now compute the sine and cosine numbers of these special angles.

Theorem 2.12: sin 0 = 0 and cos 0 = 1.

Proof:

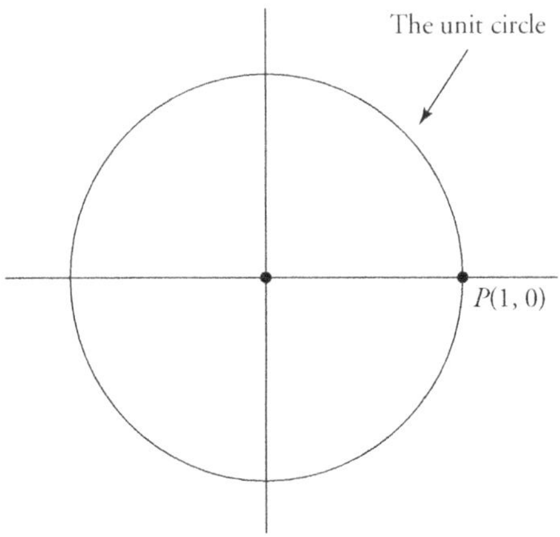

The terminal side of the angle 0 intersects the unit circle at $P(1, 0)$. Therefore, by *Theorem 2.6*, cos 0 = 1 and sin 0 = 0.

Theorem 2.13: $\sin\frac{\pi}{2} = 1$ and $\cos\frac{\pi}{2} = 0$.

Proof:

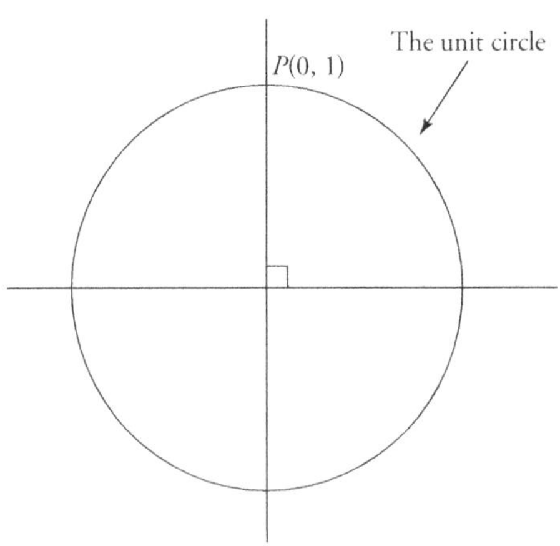

The terminal side of the angle $\frac{\pi}{2}$ intersects the unit circle at $P(0, 1)$. Therefore, by *Theorem 2.6*, $\cos\frac{\pi}{2} = 0$ and $\sin\frac{\pi}{2} = 1$.

Theorem 2.14: $\sin\frac{\pi}{4} = \frac{\sqrt{2}}{2}$ and $\cos\frac{\pi}{4} = \frac{\sqrt{2}}{2}$.

Proof:

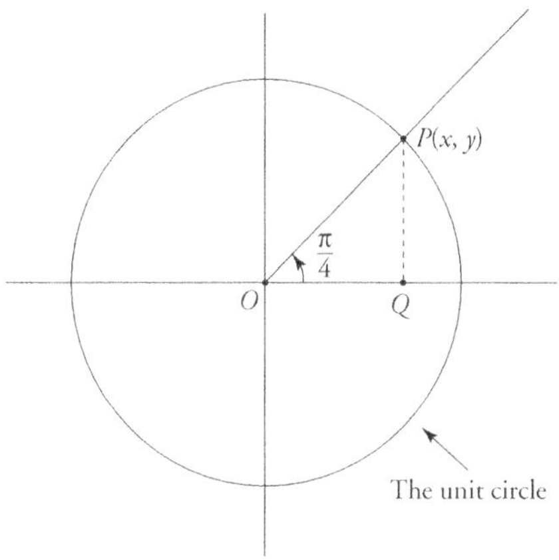

Let the point of intersection of the unit circle and the terminal point of $\frac{\pi}{4}$ be $P(x, y)$. Draw a perpendicular from point P to the x-axis and let the foot of this perpendicular be point Q. Then ΔOPQ is a right triangle. Since $\angle QOP = \frac{\pi}{4}$, angle $\angle OPQ = \frac{\pi}{4}$. Therefore, ΔOPQ is an isosceles triangle so that $OQ = PQ$. That is, $x = y$. By the Pythagorean Theorem,

$$OQ^2 + PQ^2 = OP^2$$

$$x^2 + y^2 = 1$$

$$x^2 + x^2 = 1$$

$$2x^2 = 1$$

$$x^2 = \frac{1}{2}$$

$$x = \pm\frac{1}{\sqrt{2}}$$

Since point P is in the first quadrant,

$$x = \frac{1}{\sqrt{2}} \text{ or,}$$

$$x = \frac{\sqrt{2}}{2}$$

Since $x = y$, $y = \frac{\sqrt{2}}{2}$. Therefore, $\cos\frac{\pi}{4} = \frac{\sqrt{2}}{2}$ and $\sin\frac{\pi}{4} = \frac{\sqrt{2}}{2}$.

Theorem 2.15: $\sin\frac{\pi}{6} = \frac{1}{2}$ and $\cos\frac{\pi}{6} = \frac{\sqrt{3}}{2}$.

Proof:

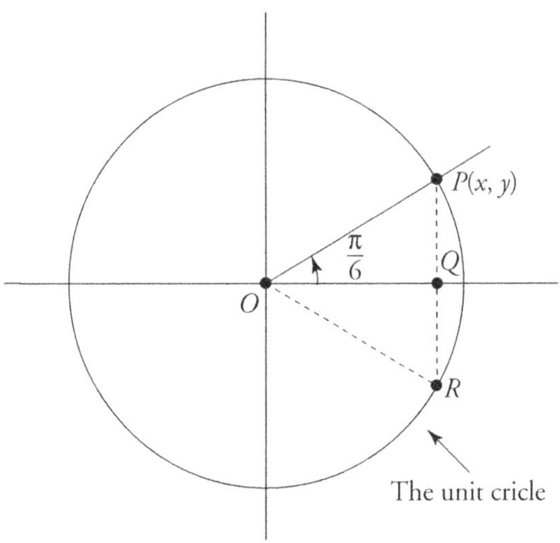

Let $P(x, y)$ be the point of intersection of the unit circle and the terminal side of the angle $\frac{\pi}{6}$. Draw a perpendicular from point P to the x-axis and let point Q be the foot of the perpendicular. Extend the perpendicular until it intersects the unit circle again at point R. Connect point R to point O by a line segment.

In the right triangle ΔOPQ, the $m\angle QOP = \frac{\pi}{6}$. Therefore, $m\angle OPQ = \frac{\pi}{3}$. ΔOPR is an isosceles triangle since $OP = OR = 1$. Therefore, $m\angle ORP = \frac{\pi}{3}$. Then, $m\angle POR = \frac{\pi}{3}$. Therefore, ΔOPR is an equilateral triangle and $PR = OP = 1$.

Line $\overleftrightarrow{OQ}$ is the perpendicular bisector of segment $\overline{PR}$. Therefore, $PQ = \frac{1}{2}$ and $y = \frac{1}{2}$.

By the Pythagorean Theorem,

$$OQ^2 + PQ^2 = OP^2$$

$$x^2 + \left(\frac{1}{2}\right)^2 = 1$$

$$x^2 + \frac{1}{4} = 1$$

$$x^2 = 1 - \frac{1}{4}$$

$$x^2 = \frac{3}{4}$$

$$x = \pm\frac{\sqrt{3}}{2}$$

Since point P is in the first quadrant, $x = \frac{\sqrt{3}}{2}$.

Therefore, $\cos\frac{\pi}{6} = \frac{\sqrt{3}}{2}$ and $\sin\frac{\pi}{6} = \frac{1}{2}$.

Theorem 2.16: $\sin\frac{\pi}{3} = \frac{\sqrt{3}}{2}$ and $\cos\frac{\pi}{3} = \frac{1}{2}$.

Proof: Let $P(x, y)$ be the point of intersection of the unit circle and the terminal side of the angle $\frac{\pi}{3}$. Draw a perpendicular from point P to the x-axis and let point Q be the foot of the perpendicular. Let point R be the point of intersection of the initial side of $\frac{\pi}{3}$ and the unit circle. Connect points P and R with a line segment.

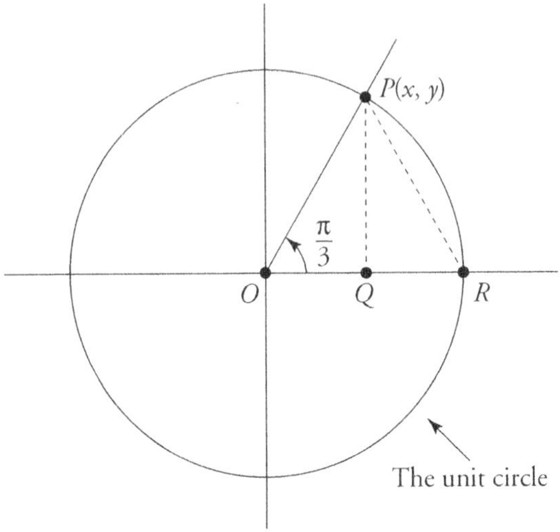

ΔOPR is an isosceles triangle because $OP = OR = 1$. Then $m\angle OPR = m\angle ORP$. Since $m\angle ROP = \frac{\pi}{3}$, both $m\angle OPR$ and $m\angle ORP = \frac{\pi}{3}$. Therefore, ΔOPR is an equilateral triangle and line $\overleftrightarrow{PQ}$ is the perpendicular bisector of segment $\overline{OR}$. That is, $x = OQ = \frac{1}{2}OR = \frac{1}{2}$.

By using the Pythagorean Theorem on right triangle ΔOPQ,

$$PQ^2 + OQ^2 = OP^2$$

$$y^2 + \left(\frac{1}{2}\right)^2 = 1$$

$$y^2 + \frac{1}{4} = 1$$

$$y^2 = 1 - \frac{1}{4}$$

$$y^2 = \frac{3}{4}$$

$$y = \pm\frac{\sqrt{3}}{2}$$

Since point P is in the first quadrant, $y = \frac{\sqrt{3}}{2}$. Therefore, $\sin\frac{\pi}{3} = \frac{\sqrt{3}}{2}$ and $\cos\frac{\pi}{3} = \frac{1}{2}$.

Table 2-2 Sine and Cosine Values of Special Angles

θ	°	$\sin\theta$	$\cos\theta$
0	0°	0	1
$\frac{\pi}{6}$	30°	$\frac{1}{2}$	$\frac{\sqrt{3}}{2}$
$\frac{\pi}{4}$	45°	$\frac{\sqrt{2}}{2}$	$\frac{\sqrt{2}}{2}$
$\frac{\pi}{3}$	60°	$\frac{\sqrt{3}}{2}$	$\frac{1}{2}$
$\frac{\pi}{2}$	90°	1	0

Example 11: Find the exact values of $\cos\frac{2\pi}{3}$ and $\sin\frac{2\pi}{3}$.

Since $\frac{2\pi}{3} = \pi - \frac{\pi}{3}$ by *Theorem 2.7*, $\cos\frac{2\pi}{3} = \cos\left(\pi - \frac{\pi}{3}\right)$ and $\sin\frac{2\pi}{3} = \sin\left(\pi - \frac{\pi}{3}\right)$.

By *Theorem 2.11*, $\sin\left(\pi - \frac{\pi}{3}\right) = \sin\frac{\pi}{3}$ and $\cos\left(\pi - \frac{\pi}{3}\right) = -\cos\frac{\pi}{3}$.

By *Theorem 2.16*, $\sin\frac{\pi}{3} = \frac{\sqrt{3}}{2}$ and $\cos\frac{\pi}{3} = \frac{1}{2}$.

Therefore, $\cos\frac{2\pi}{3} = -\frac{1}{2}$ and $\sin\frac{2\pi}{3} = \frac{\sqrt{3}}{2}$.

Example 12: Find the exact values of $\cos\frac{3\pi}{4}$ and $\sin\frac{3\pi}{4}$.

Since $\frac{3\pi}{4} = \pi - \frac{\pi}{4}$ by *Theorem 2.7,* $\cos\frac{3\pi}{4} = \cos\left(\pi - \frac{\pi}{4}\right)$ and $\sin\frac{3\pi}{4} = \sin\left(\pi - \frac{\pi}{4}\right)$.

By *Theorem 2.11,* $\sin\left(\pi - \frac{\pi}{4}\right) = \sin\frac{\pi}{4}$ and $\cos\left(\pi - \frac{\pi}{4}\right) = -\cos\frac{\pi}{4}$.

By *Theorem 2.14,* $\sin\frac{\pi}{4} = \frac{\sqrt{2}}{2}$ and $\cos\frac{\pi}{4} = \frac{\sqrt{2}}{2}$.

Therefore, $\cos\frac{3\pi}{4} = -\frac{\sqrt{2}}{2}$ and $\sin\frac{3\pi}{4} = \frac{\sqrt{2}}{2}$.

Example 13: Find the exact values of $\cos\frac{5\pi}{6}$ and $\sin\frac{5\pi}{6}$.

Since $\frac{5\pi}{6} = \pi - \frac{\pi}{6}$ by *Theorem 2.7,* $\cos\frac{5\pi}{6} = \cos\left(\pi - \frac{\pi}{6}\right)$ and $\sin\frac{5\pi}{6} = \sin\left(\pi - \frac{\pi}{6}\right)$.

By *Theorem 2.11,* $\sin\left(\pi - \frac{\pi}{6}\right) = \sin\frac{\pi}{6}$ and $\cos\left(\pi - \frac{\pi}{6}\right) = -\cos\frac{\pi}{6}$.

By *Theorem 2.15,* $\sin\frac{\pi}{6} = \frac{1}{2}$ and $\cos\frac{\pi}{6} = \frac{\sqrt{3}}{6}$.

Therefore, $\cos\frac{5\pi}{6} = -\frac{\sqrt{3}}{2}$ and $\sin\frac{5\pi}{6} = \frac{1}{2}$.

Example 14: Find the exact values of $\cos\pi$ and $\sin\pi$.

The terminal side of the angle π intersects the unit circle at the point $(-1, 0)$. Therefore, $\cos\pi = -1$ and $\sin\pi = 0$.

By using *Theorems 2.6, 2.7,* and *2.10* and the sine and cosine numbers of known angles (*Theorems 2.12–2.16* and Examples 9–12), we can find sine and cosine numbers of the angles $\frac{7\pi}{6}, \frac{5\pi}{4}, \frac{4\pi}{3}, \frac{3\pi}{2}, \frac{5\pi}{3}, \frac{7\pi}{4}, \frac{11\pi}{6}$, and 2π. Examples 15 and 16 show this process for angles $\frac{7\pi}{6}$ and $\frac{5\pi}{3}$.

Example 15: Find the exact values of $\cos\frac{7\pi}{6}$ and $\sin\frac{7\pi}{6}$.

Since $\frac{7\pi}{6}$ and $-\frac{5\pi}{6}$ are coterminal angles, by *Theorem 2.7* $\cos\frac{7\pi}{6} = \cos\left(-\frac{5\pi}{6}\right)$ and $\sin\frac{7\pi}{6} = \sin\left(-\frac{5\pi}{6}\right)$.

By *Theorem 2.10,* $\cos\left(-\frac{5\pi}{6}\right) = \cos\frac{5\pi}{6}$ and $\sin\left(-\frac{5\pi}{6}\right) = -\sin\frac{5\pi}{6}$.

Then by Example 11, $\cos\frac{7\pi}{6} = -\frac{\sqrt{3}}{2}$ and $\sin\frac{7\pi}{6} = -\frac{1}{2}$.

Example 16: Find the exact values of $\cos\frac{5\pi}{3}$ and $\sin\frac{5\pi}{3}$.

Since $\frac{5\pi}{3}$ and $-\frac{\pi}{3}$ are coterminal angles, by *Theorem 2.7* $\cos\frac{5\pi}{3} = \cos\left(-\frac{\pi}{3}\right)$ and $\sin\frac{5\pi}{3} = \sin\left(-\frac{\pi}{3}\right)$.

By *Theorem 2.10,* $\cos\left(-\frac{\pi}{3}\right) = \cos\frac{\pi}{3}$ and $\sin\left(-\frac{\pi}{3}\right) = -\sin\frac{\pi}{3}$.

Then by Example 11, $\cos\frac{5\pi}{3} = \frac{1}{2}$ and $\sin\frac{5\pi}{3} = -\frac{\sqrt{3}}{2}$.

The table below is an update of Table 2-2 to include all of these angles.

Sine and Cosine Values of Known Angles

θ	°	*sin* θ	*cos* θ
0	0°	0	1
$\frac{\pi}{6}$	30°	$\frac{1}{2}$	$\frac{\sqrt{3}}{2}$
$\frac{\pi}{4}$	45°	$\frac{\sqrt{2}}{2}$	$\frac{\sqrt{2}}{2}$

(Continued)

θ	°	$\sin\theta$	$\cos\theta$
$\frac{\pi}{3}$	60°	$\frac{\sqrt{3}}{2}$	$\frac{1}{2}$
$\frac{\pi}{2}$	90°	1	0
$\frac{2\pi}{3}$	120°	$\frac{\sqrt{3}}{2}$	$-\frac{1}{2}$
$\frac{3\pi}{4}$	135°	$\frac{\sqrt{2}}{2}$	$-\frac{\sqrt{2}}{2}$
$\frac{5\pi}{6}$	150°	$\frac{1}{2}$	$-\frac{\sqrt{3}}{2}$
π	180°	0	–1
$\frac{7\pi}{6}$	210°	$-\frac{1}{2}$	$-\frac{\sqrt{3}}{2}$
$\frac{5\pi}{4}$	225°	$-\frac{\sqrt{2}}{2}$	$-\frac{\sqrt{2}}{2}$
$\frac{4\pi}{3}$	240°	$-\frac{\sqrt{3}}{2}$	$-\frac{1}{2}$
$\frac{3\pi}{2}$	270°	–1	0
$\frac{5\pi}{3}$	300°	$-\frac{\sqrt{3}}{2}$	$\frac{1}{2}$
$\frac{7\pi}{4}$	315°	$-\frac{\sqrt{2}}{2}$	$\frac{\sqrt{2}}{2}$
$\frac{11\pi}{6}$	330°	$-\frac{1}{2}$	$\frac{\sqrt{3}}{2}$
2π	360°	0	1

Tangent, Cosecant, Secant, and Cotangent Numbers of Angles

We will define four new numbers of a given angle θ. These are known as tangent θ, written tan θ; cosecant θ, written csc θ; secant θ, written sec θ; and cotangent θ, written cot θ. The following are the definitions:

$$\tan\theta = \frac{\sin\theta}{\cos\theta}, \quad \csc\theta = \frac{1}{\sin\theta}, \quad \sec\theta = \frac{1}{\cos\theta}, \text{ and } \cot\theta = \frac{1}{\tan\theta} = \frac{\cos\theta}{\sin\theta}$$

Since all four new numbers are quotients, they exist only if denominators are not zero. Together with sin θ and cos θ, these six numbers are known as trigonometric numbers of θ.

Theorem 2.17: Consider a right triangle with an acute angle θ. Then

$$\tan\theta = \frac{\text{length of the opposite side of } \theta}{\text{length of the adjacent side of } \theta}.$$

Theorem 2.17 follows from *Theorem 2.3* and the definition of tan θ.

Theorem 2.18: Suppose θ is an angle in standard position and $P(x, y)$ is a point on the terminal side of θ other than O. Then, $\tan\theta = \frac{y}{x}$, $\csc\theta = \frac{r}{y}$, $\sec\theta = \frac{r}{x}$, and $\cot\theta = \frac{x}{y}$, if exists, where $r = OP$.

Theorem 2.18 follows from the definitions of sin θ and cos θ.

It is important to note that $r \neq 0$, since $P(x, y)$ is a point on the terminal side of θ other than O. However, either x or y can be 0. Therefore, tan θ, csc θ, sec θ, and cot θ may not exist for certain angles θ. For example, $\tan\frac{\pi}{2}$ is undefined because $\cos\frac{\pi}{2} = 0$. The following table lists the new trigonometric numbers of special angles.

Table 2-3 Tangent, Cosecant, Secant, and Cotangent Numbers of Special Angles

θ	°	tan θ	csc θ	sec θ	cot θ
0	0	0	undefined	1	undefined
$\frac{\pi}{6}$	30°	$\frac{1}{\sqrt{3}}$	2	$\frac{2}{\sqrt{3}}$	$\sqrt{3}$
$\frac{\pi}{4}$	45°	1	$\sqrt{2}$	$\sqrt{2}$	1
$\frac{\pi}{3}$	60°	$\sqrt{3}$	$\frac{2}{\sqrt{3}}$	2	$\frac{1}{\sqrt{3}}$
$\frac{\pi}{2}$	90°	undefined	1	undefined	0

Example 17: Find the six trigonometric numbers of an angle α that is in standard position and whose terminal side passes through point (–5, 12).

Point $P(-5, 12)$ is in the second quadrant. Therefore, the terminal side of α is in the second quadrant.

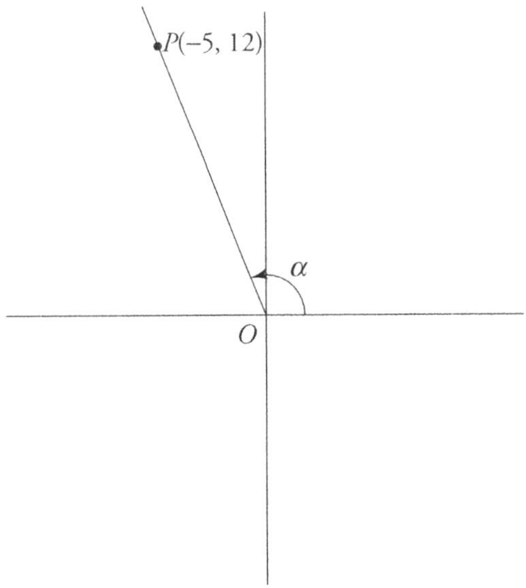

By the distance formula,

$$OP = r = \sqrt{(-5)^2 + 12^2} = \sqrt{169} = 13$$

By definitions, $\sin\alpha = \frac{12}{13}$, $\cos\alpha = -\frac{5}{13}$, $\tan\alpha = -\frac{12}{5}$, $\csc\alpha = \frac{13}{12}$, $\sec\alpha = -\frac{13}{5}$, and $\cot\alpha = -\frac{5}{12}$.

Example 18: If $\sin\theta = \frac{1}{3}$ and $\cos\theta$ is negative, then find the other five trigonometric numbers of θ.

Since $\sin\theta > 0$ and $\cos\theta < 0$, the angle θ is in the second quadrant.

$$\cos^2\theta + \sin^2\theta = 1$$

$$\cos^2\theta + \left(\frac{1}{3}\right)^2 = 1$$

$$\cos^2\theta = 1 - \frac{1}{9}$$

$$\cos^2\theta = \frac{8}{9}$$

$$\cos\theta = \pm\frac{2\sqrt{2}}{3}$$

Since cos $\theta < 0$, $\cos\theta = -\frac{2\sqrt{2}}{3}$. We can use the definitions of remaining trigonometric numbers of θ to get them.

$$\tan\theta = -\frac{1}{2\sqrt{2}},\ \ \csc\theta = 3,\ \ \sec\theta = -\frac{3}{2\sqrt{2}},\ \ \cot\theta = -2\sqrt{2}$$

Example 19: What are the exact values of sine, cosine, and tangent of 330°?

Because 330° is in the fourth quadrant, sin 330° and tan 330° are negative and cos 330° is positive. Angles 330° and –30° are coterminal, and we can use *Theorems 2.10* and *2.15* to get the required number.

$$\sin 330^\circ = \sin(-30^\circ) = -\sin 30^\circ = -\frac{1}{2}$$

$$\cos 330^\circ = \cos(-30^\circ) = \cos 30^\circ = \frac{\sqrt{3}}{2}$$

By the definition of tan 330°,

$$\tan 330^\circ = \frac{\sin 330^\circ}{\cos 330^\circ} = -\frac{1}{\sqrt{3}}$$

Example 20: If the angle θ is in the second quadrant and $\tan\theta = -\frac{4}{3}$, then find cos θ.

Since the angle θ is in the second quadrant, any point P on the terminal side other than the origin O has a negative x-coordinate and a positive y-coordinate. Since $\tan\theta = -\frac{4}{3}$, the point $P(-3, 4)$ is on the terminal side of θ.

By the distance formula, $OP = r = \sqrt{(-3)^2 + 4^2} = \sqrt{25} = 5$. Therefore, $\cos\theta = -\frac{3}{5}$.

The following theorem follows from the definitions of the tangent and the cotangent numbers of an angle θ.

Theorem 2.19: For a given angle θ, $\cot\theta = \frac{\cos\theta}{\sin\theta}$ if $\sin\theta \neq 0$.

Chapter Check-Out

Questions

1. True or false: The angle –20° is coterminal with the angle 700°.
2. True or false: If the angle θ is in the third quadrant, then tan θ is positive.
3. If $0° < \theta < 90°$ and $\sin\theta = \frac{2}{3}$, find cos θ.
4. If $0° < \theta < 90°$ and $\tan\theta = \frac{4}{3}$, find sin θ.
5. What is the exact value of cos 210°?
6. What is the exact value of sin 390°?
7. What is the exact value of $\tan\frac{5\pi}{4}$?

Answers

1. True
2. True
3. $\cos\theta = \frac{\sqrt{5}}{3}$
4. $\sin\theta = \frac{4}{5}$
5. $-\frac{\sqrt{3}}{2}$
6. $\frac{1}{2}$
7. 1

Chapter 3

TRIGONOMETRIC FUNCTIONS

Chapter Check-In

- ❑ Defining trigonometric functions
- ❑ Sketching graphs of trigonometric functions
- ❑ Identifying period, amplitude, and phase shift of a trigonometric function
- ❑ Figuring out the symmetries of trigonometric functions
- ❑ Sketching the graphs of general trigonometric functions

Common Core Standard: Extend the domain of trigonometric functions using the unit circle

Explain how the unit circle in the coordinate plane enables the extension of trigonometric functions to all real numbers, interpreted as radian measures of angles traversed counterclockwise around the unit circle. (HSF.TF.A.2)

Use special triangles to determine geometrically the values of sine, cosine, and tangent for $\frac{\pi}{3}$, $\frac{\pi}{4}$, and $\frac{\pi}{6}$, and use the unit circle to express the values of sine, cosine, and tangent for x, $\pi + x$, and $2\pi - x$ in terms of their values for x, where x is any real number. (HSF.TF.A.3)

Use the unit circle to explain symmetry (odd and even) and periodicity of trigonometric functions. (HSF.TF.A.4)

Choose trigonometric functions to model periodic phenomena with specified amplitude, frequency, and midline. (HSF.TF.B.5)

Please also refer to the standards for mathematical practice, found here: www.corestandards.org/Math/Practice/.

We have defined trigonometric numbers of angles and we know how to find those numbers for special angles and multiples of special angles. We will learn later how to find exact values of trigonometric numbers of a few more angles, but for most angles, the best we can do is to find them approximately. There are calculus methods to find trigonometric numbers accurately up to any given decimal place. Any good calculator is programmed to calculate trigonometric numbers, usually for up to ten decimal place accuracy. This is only possible if we look at the *whole collection of all such numbers* for all angles.

Trigonometry Theorems

The trigonometry theorems in this book are numbered for organizational purposes. For example, *Theorem 2.2* is the second theorem given in Chapter 2. Study these theorems by their content, not by their number, as the theorem numbering has no significance outside of this book.

Sine Function

Imagine looking at *all* sine numbers of *all* angles together. One thing we notice is that for each angle, there is exactly one sine number. Such collections are known as *functions*. More precisely, the collection of all ordered pairs (θ, sin θ), where θ is an angle, is a function of θ, because for each θ, there is exactly one sin θ. By using the conventional notation, that is, $f(\theta)$, pronounced "f of θ," we can write the sine function as $f(\theta) = \sin \theta$. When you use this notation, it is understood that you are looking at the entire collection of ordered pairs (θ, sin θ) or (θ, $f(\theta)$). The sine number of a particular angle can be obtained in this notation by replacing θ with the particular angle. For example, $f\left(\frac{\pi}{6}\right) = \frac{1}{2}$. However, when looking at the last equation, it is not clear what the function is. Therefore, usually the letter f is replaced with the name of the function when writing function values. That is, instead of writing $f\left(\frac{\pi}{6}\right) = \frac{1}{2}$, we would write $\sin\left(\frac{\pi}{6}\right) = \frac{1}{2}$. Conventionally, the parentheses are dropped and this equation is written as $\sin\frac{\pi}{6} = \frac{1}{2}$. Therefore, from now on in this book, sin θ refers to the value of the sine function at θ.

We can represent the sine function graphically by looking at each ordered pair (θ, sin θ), for any given θ as a point on the Cartesian coordinate system with the x-axis replaced with the θ-axis and the y-axis replaced with the sin θ-axis. The following figure shows the location of the point $\left(\frac{\pi}{6}, \sin\frac{\pi}{6}\right)$ on this coordinate system.

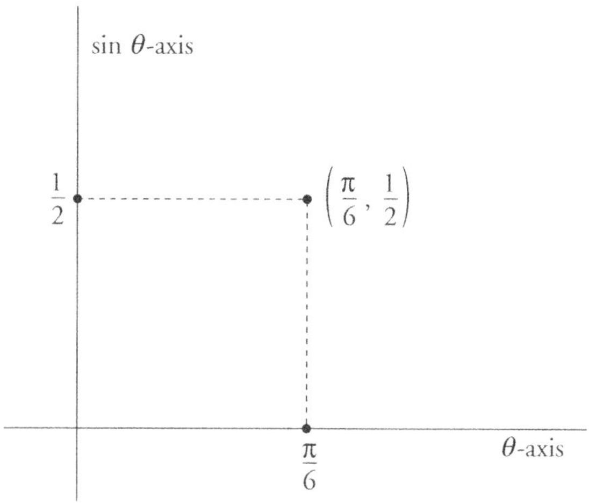

The collection of all points (θ, sin θ) for all θ is called the graph of the function $f(\theta)$ = sin θ. We know that the y-coordinate of the point of intersection of the terminal side of any angle θ in standard position is sin θ.

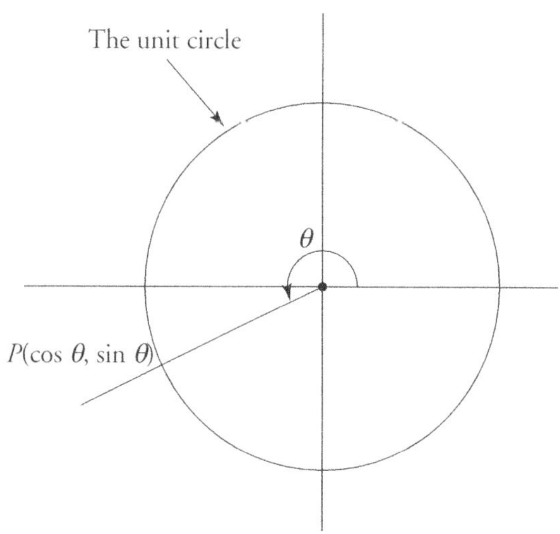

As a result, we observed that sin $(2n\pi + \theta) = \sin\theta$, for any θ and any integer n (*Theorem 2.9*). If we can plot the graph of the sine function for the interval $[0, 2\pi]$, then *Theorem 2.9* indicates that the graph of sine function in $[0, 2\pi]$ is identical to the graph of sine function in the intervals $[2\pi, 4\pi]$, $[4\pi, 6\pi]$, etc. *Theorem 2.9* is also true for negative integers, and therefore, the graph of sine function in $[0, 2\pi]$ is identical to the graph of sine function in $[-2\pi, 0]$, $[-4\pi, -2\pi]$, etc. Therefore, if we can get the graph of sine function in the interval $[0, 2\pi]$, then we know the graph of the entire sine function. We can use the exact values of $\sin\theta$ that we obtained in Chapter 2; see pp. 37–38 for special angles and their multiples to plot the graph of $f(\theta) = \sin\theta$ in $[0, 2\pi]$. The following is a sketch of the graph of sine function.

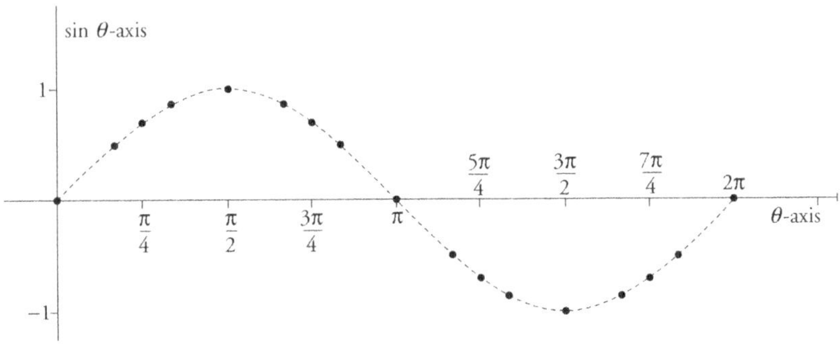

Sine function is an example of a periodic function. A function $f(\theta)$ is called a **periodic function** if there is a number p so that $f(\theta + p) = f(\theta)$, for all θ in the domain of the function. The smallest of such p is called the **period** of f. The sine function is periodic because $\sin(2n\pi + \theta) = \sin\theta$, for any θ, by *Theorem 2.9*. The period of the sine function is 2π. The graph of the sine function in the interval $[0, 2\pi]$ is called one *cycle* of the sine function. This cycle repeats in intervals $[2\pi, 4\pi]$, $[4\pi, 6\pi]$, etc. and $[-2\pi, 0]$, $[-4\pi, -2\pi]$, etc. The length of any of these intervals is the period of $f(\theta) = \sin\theta$. That is, the period is 2π.

Theorem 3.1: For any angle θ, $-1 \leq \sin\theta \leq 1$.

Proof: For any given θ, the y-coordinate of the point of intersection of the terminal side of θ and the unit circle is $\sin\theta$. Since the y-coordinate of any point on the unit circle lies between 1 and −1, inclusive, *Theorem 3.1* follows.

We say the **maximum value** of the sine function is 1 and the **minimum value** of the sine function is −1. As you can see from the graph of the sine function below, it has a maximum value at $\theta = \frac{\pi}{2}$ and a minimum value at $\theta = \frac{3\pi}{2}$ in the interval $[0, 2\pi]$. The number $\left|\frac{(\text{maximum value} - \text{minimum value})}{2}\right|$ is called the **amplitude of a periodic function.** Therefore, the amplitude of the sine function is 1.

The angles θ where $f(\theta) = 0$ are called *zeros* of the function f. The zeros of the sine function are 0, π, and 2π in the interval $[0, 2\pi]$. The following cycle of the graph of the sine function indicates only those significant angles.

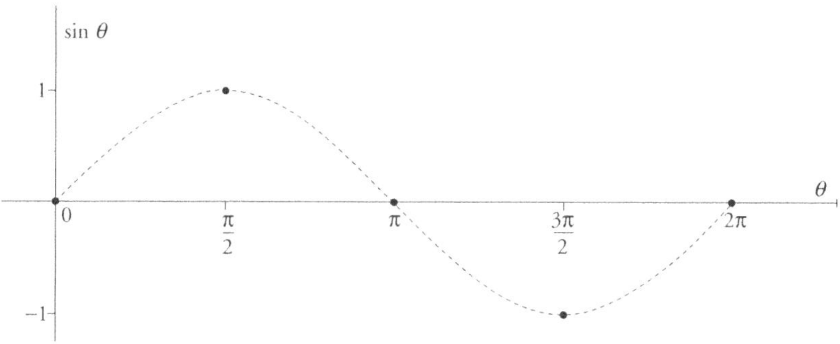

The collection of all possible angles, where the sine function is defined, is called the *domain* of the sine function. As you can see, there are no restrictions on angles θ where $\sin\theta$ is defined. Therefore, we say the domain of the sine function is the interval $(-\infty, \infty)$. The collection of all possible values of $\sin\theta$ for all angles θ is the *range* of the sine function. From *Theorem 3.1*, the range of the sine function is the interval $[-1, 1]$.

The following is a portion of the graph of the sine function.

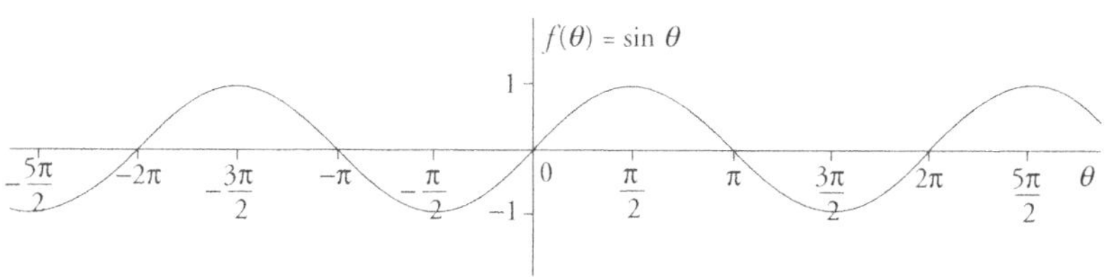

A function f that satisfies the condition $f(-x) = f(x)$ for all x is called an **even function**. If we reflect the graph of an even function across the vertical axis, then the graph remains same. We say the graph of an even function is symmetrical with respect to the vertical axis.

A function f that satisfies the condition $f(-x) = -f(x)$ for all x is called an **odd function**. This means the reflection of the graph of an odd function across the vertical axis followed by a reflection across the horizontal axis preserves the graph of the function. We say the graph of an odd function is symmetrical with respect to the origin. Since $\sin(-x) = -\sin x$, the sine function is an odd function.

Cosine Function

Theorem 3.2: For any angle θ, $-1 \leq \cos\theta \leq 1$.

Proof: For any given θ, the x-coordinate of the point of intersection of the terminal side of θ and the unit circle is $\cos\theta$. Since the x-coordinate of any point on the unit circle lies between 1 and -1, inclusive, *Theorem 3.2* follows.

By *Theorem 2.9*, $\cos(2n\pi + \theta) = \cos\theta$ for any θ for any integer n. Therefore, the period of the cosine function, $f(\theta) = \cos\theta$, is 2π. We can use the cosine numbers of special angles and their multiples in Table 2-2 to sketch one cycle of the cosine function, as shown below.

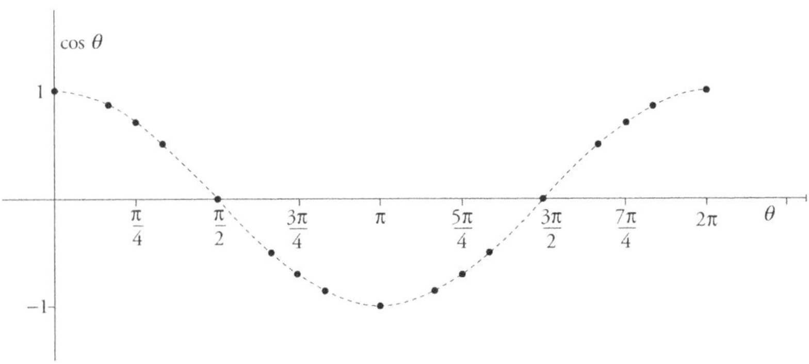

The maximum value of the cosine function is 1, and the minimum value of the cosine function is -1. As you can see from the graph of the cosine function, it has a maximum value at $\theta = 0$ and at $\theta = 2\pi$ in the interval $[0, 2\pi]$. The cosine function has a minimum value at $\theta = \pi$ in the interval $[0, 2\pi]$. The amplitude of the cosine function is 1, and the zeros of the cosine function in the interval $[0, 2\pi]$ are $\frac{\pi}{2}$ and $\frac{3\pi}{2}$. The following cycle of the graph of the cosine function indicates only those significant angles, where the function has zeros and extreme values.

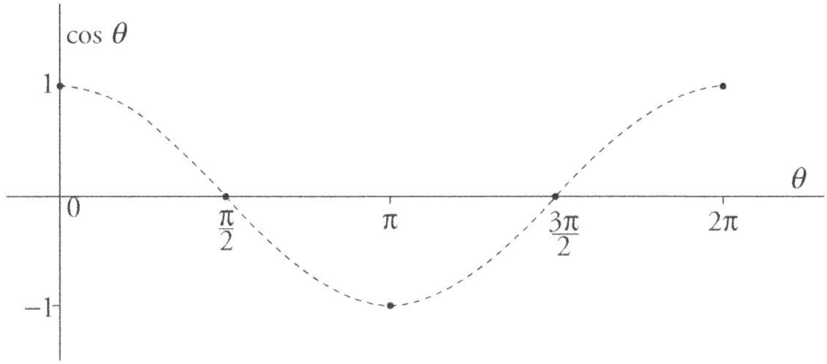

The domain of the cosine function is $(-\infty, \infty)$ and the range of the cosine function is $[-1, 1]$.

The following is a portion of the graph of the cosine function.

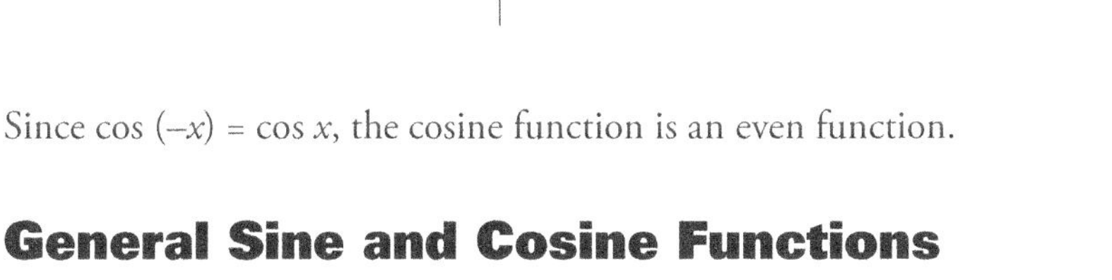

Since $\cos(-x) = \cos x$, the cosine function is an even function.

General Sine and Cosine Functions

When angles are measured in radians, the measures are real numbers. The domains of these functions are all real numbers. Therefore, without loss of any generality, we will replace the letter θ with x and think of trigonometric functions as functions of x. That is, we will rename the θ-axis as the x-axis. We will investigate functions of the form $f(x) = A \sin B(x - C)$ or $f(x) = A \cos B(x - C)$, where A and B are non-zero constants and C is a constant. When $A = 1$, $B = 1$, and $C = 0$, we get the familiar sine and cosine functions.

We use what we know about sine and cosine functions to sketch the graphs of general sine and cosine functions.

Example 1: Sketch one cycle of the graph of $f(x) = 3 \sin 2x$ and find the amplitude and the period.

Let $\alpha = 2x$. We know how to sketch one cycle of the graph of $f(\alpha) = \sin \alpha$. In the following table, we list the significant values of $f(\alpha) = \sin \alpha$, where $f(\alpha)$ has zeros, maximum values, and minimum values.

α	$\sin \alpha$
0	0
$\frac{\pi}{2}$	1
π	0
$\frac{3\pi}{2}$	–1
2π	0

Since $\alpha = 2x$, we get $x = \frac{\alpha}{2}$. Now we will add a column for the corresponding values of x to the table above.

α	$\sin \alpha$	x
0	0	0
$\frac{\pi}{2}$	1	$\frac{\pi}{4}$
π	0	$\frac{\pi}{2}$
$\frac{3\pi}{2}$	–1	$\frac{3\pi}{4}$
2π	0	π

Finally, we add another column for $f(x) = 3 \sin 2x = 3 \sin \alpha$ and calculate the values of $f(x)$ as they correspond to the values of x using the values in the second column.

α	$\sin \alpha$	x	$f(x) = 3 \sin \alpha$
0	0	0	0
$\frac{\pi}{2}$	1	$\frac{\pi}{4}$	3
π	0	$\frac{\pi}{2}$	0
$\frac{3\pi}{2}$	–1	$\frac{3\pi}{4}$	–3
2π	0	π	0

Now we will sketch one cycle of the graph using the significant values in the last two columns of the previous table.

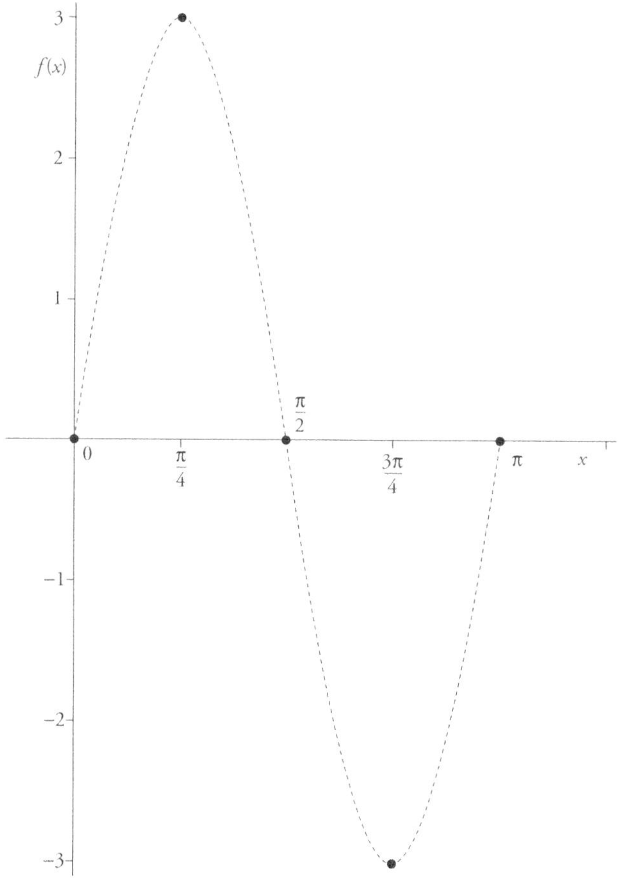

Clearly, the period of this function is π and the amplitude is 3.

Example 2: Sketch one cycle of the graph of $f(x) = 2\cos\frac{1}{2}x$ and find the amplitude and the period.

Let $\alpha = \frac{1}{2}x$. Then $x = 2\alpha$. We will construct a table as we did in Example 1.

α	$\cos\alpha$	x	$f(x) = 2\cos\alpha$
0	1	0	2
$\frac{\pi}{2}$	0	π	0
π	–1	2π	–2
$\frac{3\pi}{2}$	0	3π	0
2π	1	4π	2

The following is one cycle of $f(x) = 2\cos\frac{1}{2}x$.

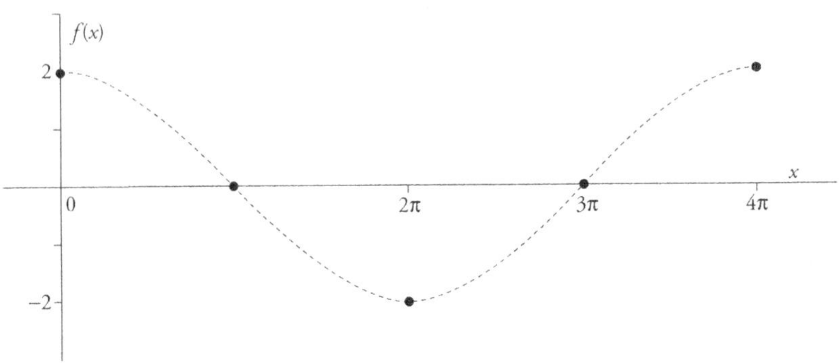

The period of this function is 4π and the amplitude is 2.

Example 3: Sketch one cycle of the graph of $f(x) = 3\sin(-2x)$ and find the amplitude and the period.

Recall from *Theorem 2.10* that $\sin(-\theta) = -\sin\theta$. Therefore, the given function can be written as $f(x) = -3\sin(2x)$.

Let $\alpha = 2x$. Then $x = \frac{\alpha}{2}$.

α	$\sin\alpha$	x	$f(x) = -3\sin\alpha$
0	0	0	0
$\frac{\pi}{2}$	1	$\frac{\pi}{4}$	–3
π	0	$\frac{\pi}{2}$	0
$\frac{3\pi}{2}$	–1	$\frac{3\pi}{4}$	3
2π	0	π	0

The following is one cycle of $f(x) = -3\sin(2x)$.

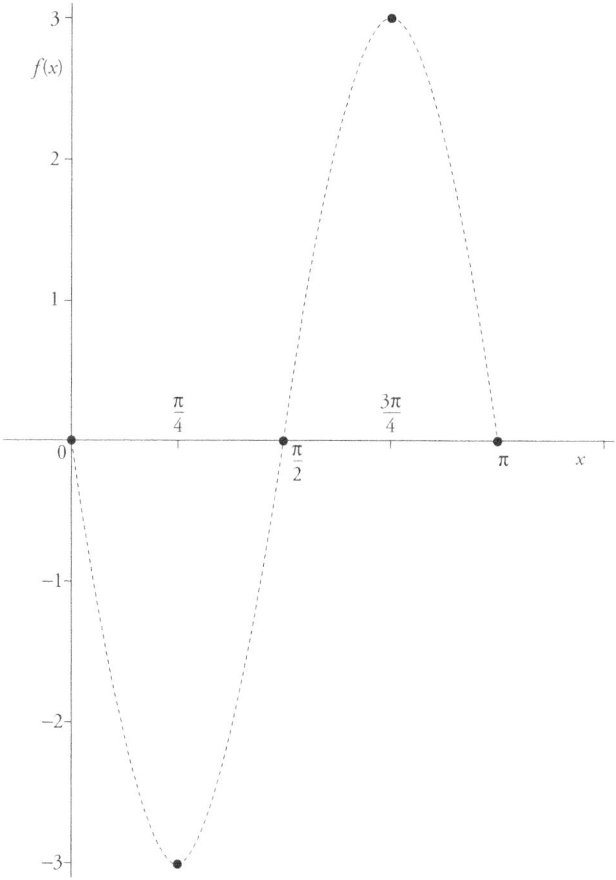

Clearly, the graph of $f(x) = -3\sin(2x)$ is the reflection of the graph of $g(x) = 3\sin(2x)$ across the x-axis.

Another number associated with the graph of a trigonometric function is the phase shift. The **phase shift** is a signed number that indicates the horizontal shift of a graph of a trigonometric function. If the shift is to the right, then the sign of the phase shift is positive; if the shift is to the left, then the sign is negative. There is no phase shift for the functions in Examples 1, 2, and 3. We could say the phase shift is 0 for those functions.

Example 4: Sketch one cycle of the graph of $f(x) = 2\sin 3\left(x - \frac{\pi}{4}\right)$ and find the amplitude, the period, and the phase shift.

Let $\alpha = 3\left(x - \frac{\pi}{4}\right)$. Then $x = \frac{\alpha}{3} + \frac{\pi}{4}$.

α	$\sin\alpha$	x	$f(x) = 2\sin\alpha$
0	0	$\frac{\pi}{4}$	0
$\frac{\pi}{2}$	1	$\frac{5\pi}{12}$	2
π	0	$\frac{7\pi}{12}$	0
$\frac{3\pi}{2}$	–1	$\frac{3\pi}{4}$	–2
2π	0	$\frac{11\pi}{12}$	0

The following is one cycle of $f(x) = 2\sin 3\left(x - \frac{\pi}{4}\right)$.

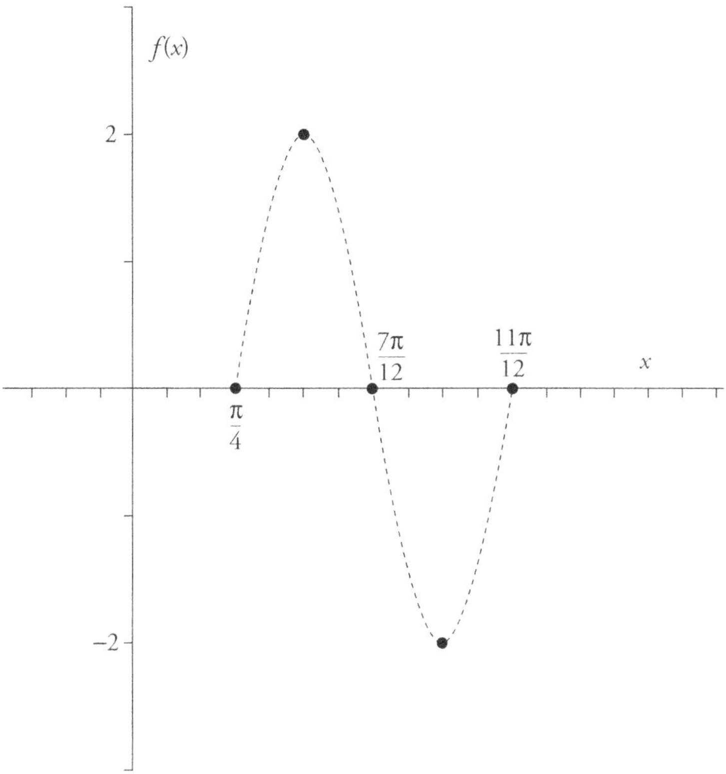

The period of this function is $\frac{11\pi}{12} - \frac{\pi}{4} = \frac{2\pi}{3}$. The amplitude is 2 and the phase shift is $+\frac{\pi}{4}$.

Example 5: Sketch one cycle of the graph of $f(x) = 3\cos 2\left(x + \frac{\pi}{3}\right)$ and find the amplitude, the period, and the phase shift.

Let $\alpha = 2\left(x + \frac{\pi}{3}\right)$. Then $x = \frac{\alpha}{2} - \frac{\pi}{3}$.

α	$\cos\alpha$	x	$f(x) = 3\cos\alpha$
0	1	$-\frac{\pi}{3}$	3
$\frac{\pi}{2}$	0	$-\frac{\pi}{12}$	0
π	–1	$\frac{\pi}{6}$	–3
$\frac{3\pi}{2}$	0	$\frac{5\pi}{12}$	0
2π	1	$\frac{2\pi}{3}$	3

The following is one cycle of $f(x) = 3\cos 2\left(x + \frac{\pi}{3}\right)$.

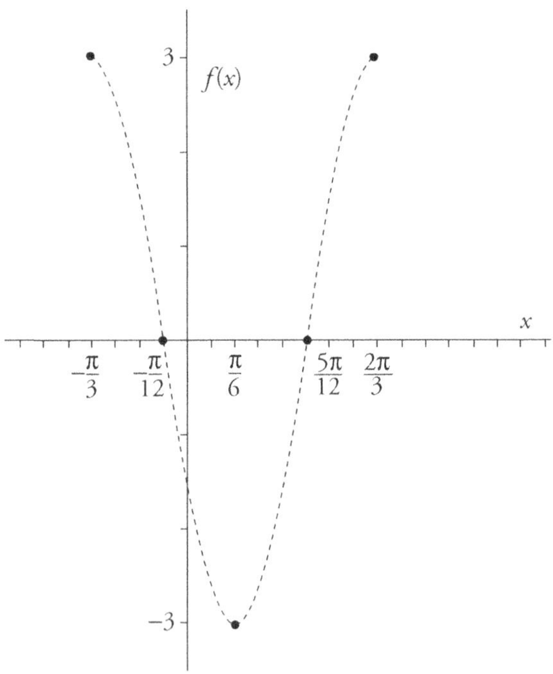

The period of this function is $\frac{2\pi}{3} - \left(-\frac{\pi}{3}\right) = \pi$. The amplitude is 3 and the phase shift is $-\frac{\pi}{3}$.

Example 6: Sketch one cycle of the graph of $f(x) = -3\sin\left[-2\left(x+\frac{\pi}{3}\right)\right]$ and find the amplitude, the period, and the phase shift.

Let $\alpha = -2\left(x+\frac{\pi}{3}\right)$. Then $x = -\frac{\alpha}{2} - \frac{\pi}{3}$.

α	sin α	x	f(x) = –3 sin α
0	0	$-\frac{\pi}{3}$	0
$\frac{\pi}{2}$	1	$-\frac{7\pi}{12}$	–3
π	0	$-\frac{5\pi}{6}$	0
$\frac{3\pi}{2}$	–1	$-\frac{13\pi}{12}$	3
2π	0	$-\frac{4\pi}{3}$	0

The following is one cycle of $f(x) = -3\sin\left[-2\left(x+\frac{\pi}{3}\right)\right]$.

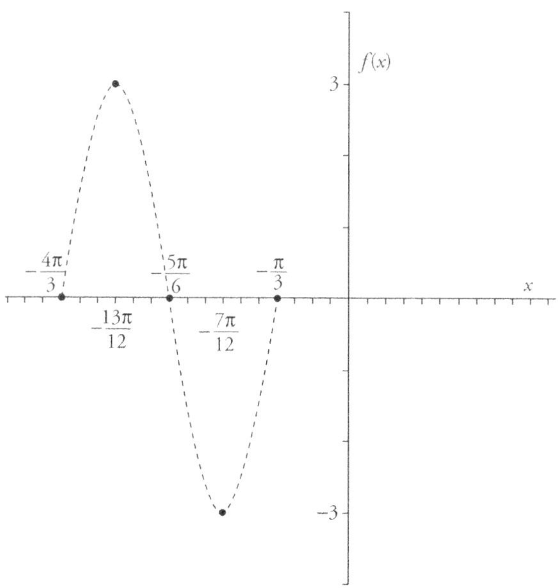

The period of this function is $-\frac{\pi}{3}-\left(-\frac{4\pi}{3}\right)=\pi$. The amplitude is 3 and the phase shift is $-\frac{\pi}{3}$.

We could do the problem in Example 6 another way by using *Theorem 2.10*. According to *Theorem 2.10*, $\sin(-\theta) = -\sin\theta$.

Therefore, $f(x)=-3\sin\left[-2\left(x+\frac{\pi}{3}\right)\right]=3\sin\left[2\left(x+\frac{\pi}{3}\right)\right]$.

Let us plot $f(x)=3\sin\left[2\left(x+\frac{\pi}{3}\right)\right]$.

Let $\alpha = 2\left(x+\frac{\pi}{3}\right)$. Then $x=\frac{\alpha}{2}-\frac{\pi}{3}$.

α	$\sin\alpha$	x	$f(x) = 3\sin\alpha$
0	0	$-\frac{\pi}{3}$	0
$\frac{\pi}{2}$	1	$-\frac{\pi}{12}$	3
π	0	$\frac{\pi}{6}$	0
$\frac{3\pi}{2}$	−1	$\frac{5\pi}{12}$	−3
2π	0	$\frac{2\pi}{3}$	0

The cycle of $f(x)=3\sin\left[2\left(x+\frac{\pi}{3}\right)\right]$ given by the above table is plotted, together with the cycle of $f(x)=-3\sin\left[-2\left(x+\frac{\pi}{3}\right)\right]$ previously plotted. The cycle of $f(x)=-3\sin\left[-2\left(x+\frac{\pi}{3}\right)\right]$ is drawn as a solid curve and the cycle of $f(x)=3\sin\left[2\left(x+\frac{\pi}{3}\right)\right]$ is drawn as a dotted curve.

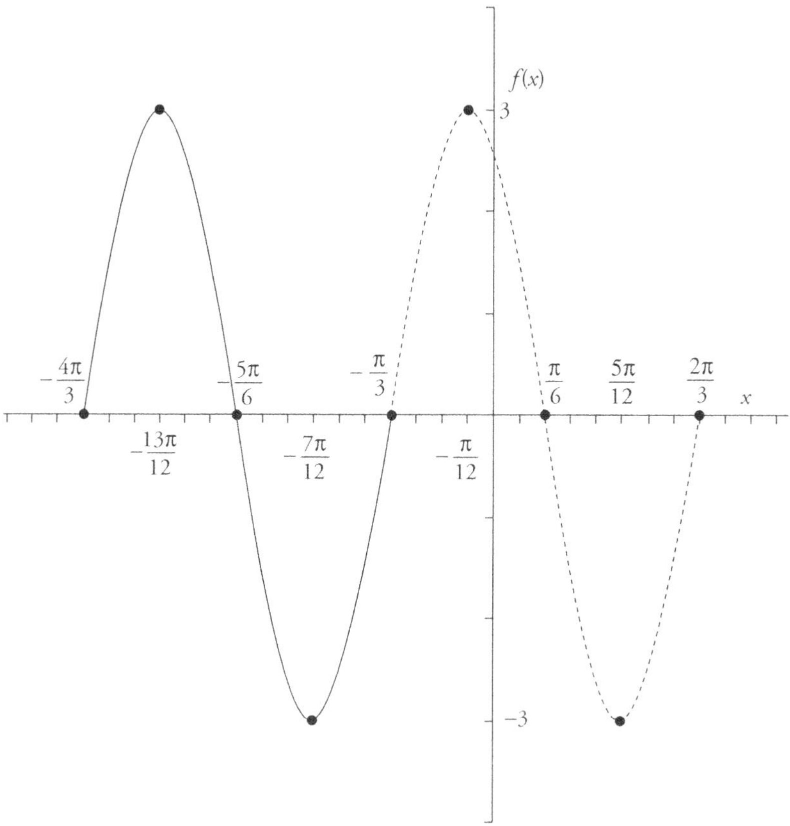

As you can see, it is the next cycle of the same periodic function.

The following two theorems summarize what you have observed in Examples 1–6.

Theorem 3.3: Suppose $f(x) = A \sin B(x - C)$, where A and B are non-zero constants and C is a constant. Then the amplitude of f is $|A|$, the period is $\frac{2\pi}{|B|}$, and the phase shift is $+C$ if $C > 0$ or $-|C|$ if $C < 0$.

Proof: Let $\alpha = B(x - C)$. Then $f(x) = A \sin \alpha$.

By *Theorem 3.1*, $-1 \le \sin \alpha \le 1$.

If A is positive, then $-A \le A \sin \alpha \le A$, and the amplitude is $\frac{(A-(-A))}{2} = A \cdot$

If A is negative, then $-A \geq A \sin \alpha \geq A$, and the amplitude is $\frac{(-A-A)}{2} = -A$. In either case, the amplitude is a positive number. Therefore, the amplitude is $|A|$.

One cycle of sin α is in the interval $[0, 2\pi]$. That is, $0 \leq \alpha \leq 2\pi$.

Suppose $B > 0$. Then

$$0 \leq \alpha \leq 2\pi$$

$$0 \leq B(x-C) \leq 2\pi$$

$$0 \leq x-C \leq \frac{2\pi}{B}$$

$$C \leq x \leq \frac{2\pi}{B}+C \quad (1)$$

Let us identify the above inequality as (1) for future reference.

That is, one cycle of f lies in the interval $\left[C,\ \frac{2\pi}{B}+C\right]$.

In this case, the period is $\frac{2\pi}{B}+C-C=\frac{2\pi}{B}$.

Now suppose $B < 0$. Then

$$0 \leq \alpha \leq 2\pi$$

$$0 \leq B(x-C) \leq 2\pi$$

$$0 \geq x-C \geq \frac{2\pi}{B}$$

$$C \geq x \geq \frac{2\pi}{B}+C \quad (2)$$

Let us identify the above inequality as (2) for future reference.

That is, one cycle of f lies in the interval $\left[\frac{2\pi}{B}+C,\ C\right]$.

In this case, the period is $C-\left(\frac{2\pi}{B}+C\right)=-\frac{2\pi}{B}$.

In either case, the period is positive and is equal to $\frac{2\pi}{|B|}$.

If C is positive and B is positive, then by (1), the phase shift of f is $+C$.

If C is positive and B is negative, then by (2), the phase shift of f is $+C$.

If C is negative and B is positive, then by (1), the phase shift of f is $C=-|C|$.

If C is negative and B is negative, then by (2), the phase shift of f is $C=-|C|$.

Theorem 3.4: Suppose $f(x) = A\cos B(x-C)$, where A and B are non-zero constants and C is a constant. Then the amplitude of f is $|A|$, the period is $\frac{2\pi}{|B|}$, and the phase shift is $+C$ if $C>0$ or $-|C|$ if $C<0$.

Proof of this theorem is similar to the proof of *Theorem 3.3*.

With *Theorems 3.3* and *3.4*, we can now sketch a graph of a general sine or cosine function, first calculating the amplitude, period, and phase shift.

Example 7: Find the amplitude, period, and phase shift of $f(x)=3\sin\left(2x-\frac{\pi}{2}\right)$ and sketch the graph of one cycle of the function.

$$f(x)=3\sin\left(2x-\frac{\pi}{2}\right)$$

$$f(x)=3\sin 2\left(x-\frac{\pi}{4}\right)$$

By *Theorem 3.3*, the amplitude is 3, the period is $\frac{2\pi}{2}=\pi$, and the phase shift is $+\frac{\pi}{4}$. The following is the sketch of one cycle of the graph.

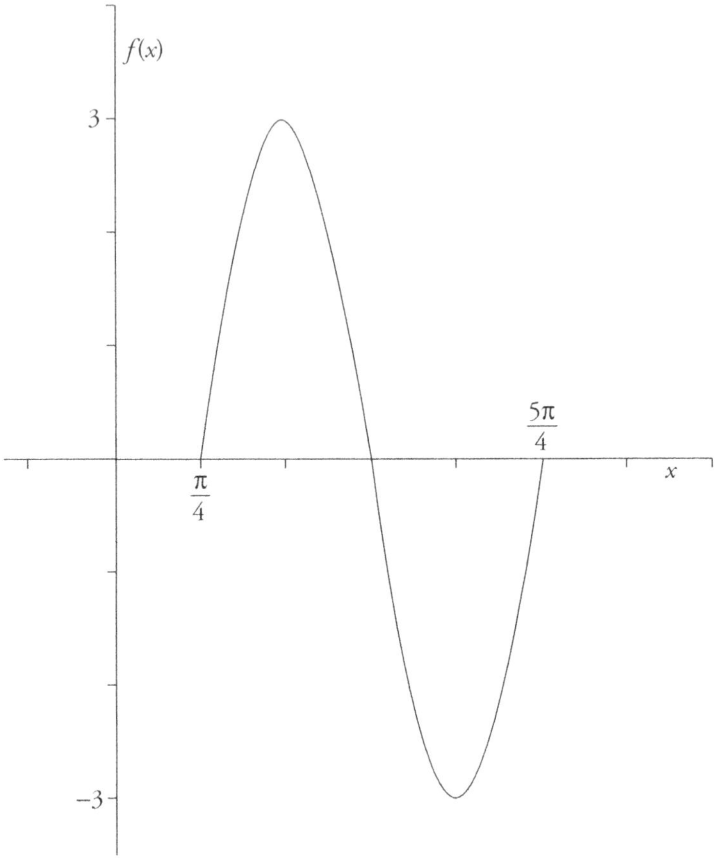

Example 8: Find the amplitude, period, and phase shift of $f(x) = \sin x + 2$ and sketch the graph of one cycle of the function.

Let $y = f(x)$. Then

$$y = \sin x + 2$$
$$y - 2 = \sin x$$

Let $y' = y - 2$. Then $y' = \sin x$. Therefore, we see the familiar sine function with the x-axis replaced by the line $y' = 0$. That is, in place of the x-axis, we have the $y = 2$ line. The amplitude is 1, the period is 2π, and the phase shift is 0. The following is one cycle of the graph of this function.

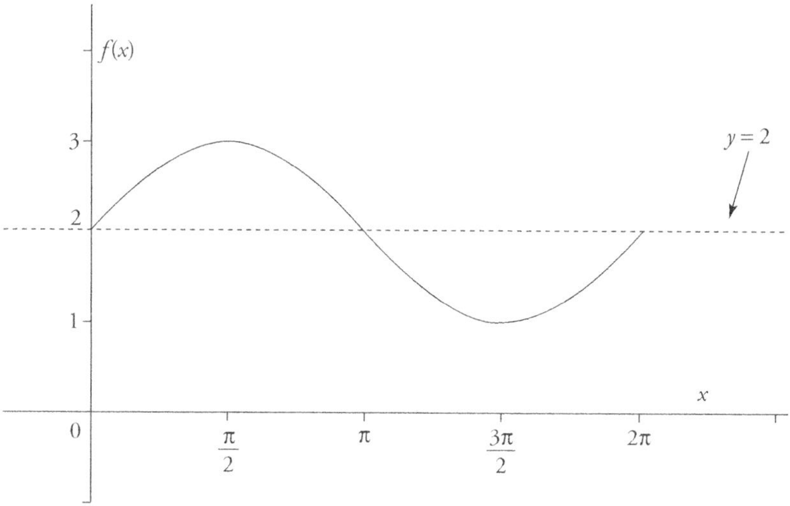

We will call the line $y = 2$; the **midline** of the function $f(x) = \sin x + 2$.

Theorem 3.5 generalizes *Theorems 3.3* and *3.4* to include this information.

Theorem 3.5: Suppose $f(x) = A \sin B(x - C) + D$, where A and B are non-zero constants and C and D are constants. Then the amplitude of f is $|A|$, the period is $\frac{2\pi}{|B|}$, the phase shift is $+C$ if $C > 0$ or $-|C|$ if $C < 0$, and the midline is $y = D$, where $y = f(x)$.

Theorem 3.6: Suppose $f(x) = A \cos B(x - C) + D$, where A and B are non-zero constants and C and D are constants. Then the amplitude of f is $|A|$, the period is $\frac{2\pi}{|B|}$, the phase shift is $+C$ if $C > 0$ or $-|C|$ if $C < 0$, and the midline is $y = D$, where $y = f(x)$.

Example 9: Sketch the graph of $y = \cos \pi x - 2$.

By *Theorem 3.5*, the amplitude is 1, the period is $\frac{2\pi}{\pi} = 2$, the phase shift is 0, and the midline is $y = -2$. The following is a sketch of one cycle of the graph.

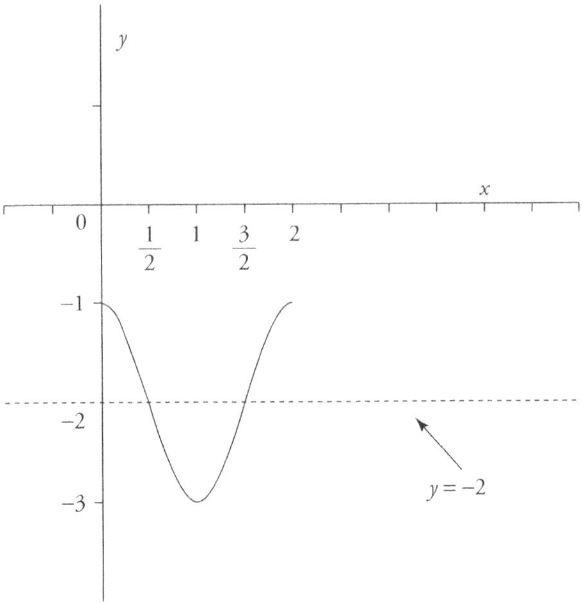

Tangent Function

By definition, $\tan x = \dfrac{\sin x}{\cos x}$. When $\cos x = 0$, the tangent function is undefined. In the interval $[0, 2\pi]$, this happens at $x = \dfrac{\pi}{2}$ and at $x = \dfrac{3\pi}{2}$.

We can get a sketch of the tangent function by looking at the graphs of the sine and cosine functions together and using tangent numbers of special angles and their multiples.

x	***tan x***	***x***	***tan x***
0	0	$\frac{7\pi}{6}$	$\frac{1}{\sqrt{3}}$
$\frac{\pi}{6}$	$\frac{1}{\sqrt{3}}$	$\frac{5\pi}{4}$	1
$\frac{\pi}{4}$	1	$\frac{4\pi}{3}$	$\sqrt{3}$
$\frac{\pi}{3}$	$\sqrt{3}$	$\frac{3\pi}{2}$	undefined
$\frac{\pi}{2}$	undefined	$\frac{5\pi}{3}$	$-\sqrt{3}$
$\frac{2\pi}{3}$	$\sqrt{3}$	$\frac{7\pi}{4}$	-1

x	$\tan x$	x	$\tan x$
$\frac{3\pi}{4}$	-1	$\frac{11\pi}{6}$	$-\frac{1}{\sqrt{3}}$
$\frac{5\pi}{6}$	$-\frac{1}{\sqrt{3}}$	2π	0
π	0		

We will use the notation $x \to a^-$ to mean that the "x is very close to the number a coming from the left of a on the x-axis"; we will use the notation $x \to a^+$ to mean that the "x is very close to the number a coming from the right of a on the x-axis"; we will use the notation $\cos x \to 0^+$ to mean that "$\cos x$ is positive but very close to 0"; we will use the notation $\cos x \to 0^-$ to mean that "$\cos x$ is negative but very close to 0"; we will use the notation $f(x) \to +\infty$ to mean that "$f(x)$ increases without bounds"; and we will use the notation $f(x) \to -\infty$ to mean that "$f(x)$ decreases without bounds."

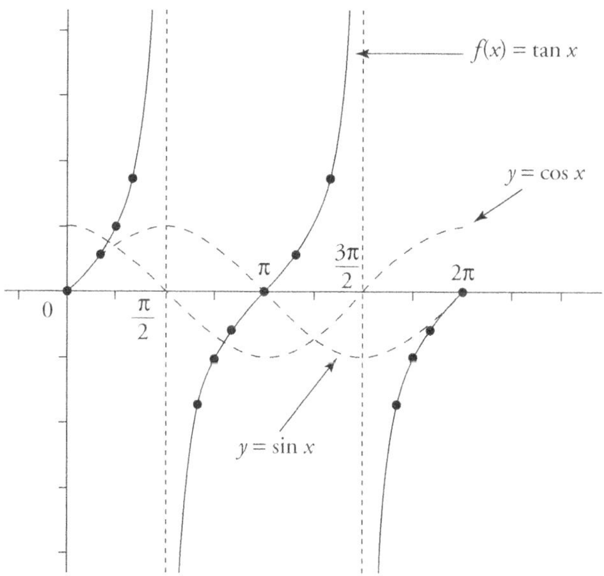

As $x \to \frac{\pi}{2}^-$, $\sin x \approx 1$ and $\cos x \to 0^+$. That is, both the numerator and the denominator of $\frac{\sin x}{\cos x}$ remain positive; the numerator remains close to 1, and the denominator gets very small. Therefore, $\tan x \to +\infty$.

As $x \to \frac{\pi^+}{2}$, $\sin x \approx 1$ and $\cos x \to 0^-$. That is, the numerator of $\frac{\sin x}{\cos x}$ remains close to 1, but the denominator is negative and gets very small. Therefore, $\tan x \to -\infty$.

We say the tangent function has a vertical asymptote at $x = \frac{\pi}{2}$.

Recall that a **vertical asymptote** of a function is a vertical line $x = a$, so that either $f(x) \to \pm\infty$ as $x \to a^-$ or $f(x) \to \pm\infty$ as $x \to a^+$.

You observe the same behavior of the tangent function near $\frac{3\pi}{2}$, as shown in the figure on p. 65.

The following figure shows the graph of the tangent function in the interval $\left(-\frac{\pi}{2}, \frac{5\pi}{2}\right)$. As you can see, the graph in this interval repeats in the intervals $\left(\frac{\pi}{2}, \frac{3\pi}{2}\right)$, $\left(\frac{3\pi}{2}, \frac{5\pi}{2}\right)$, etc. We say the tangent function is periodic with period π.

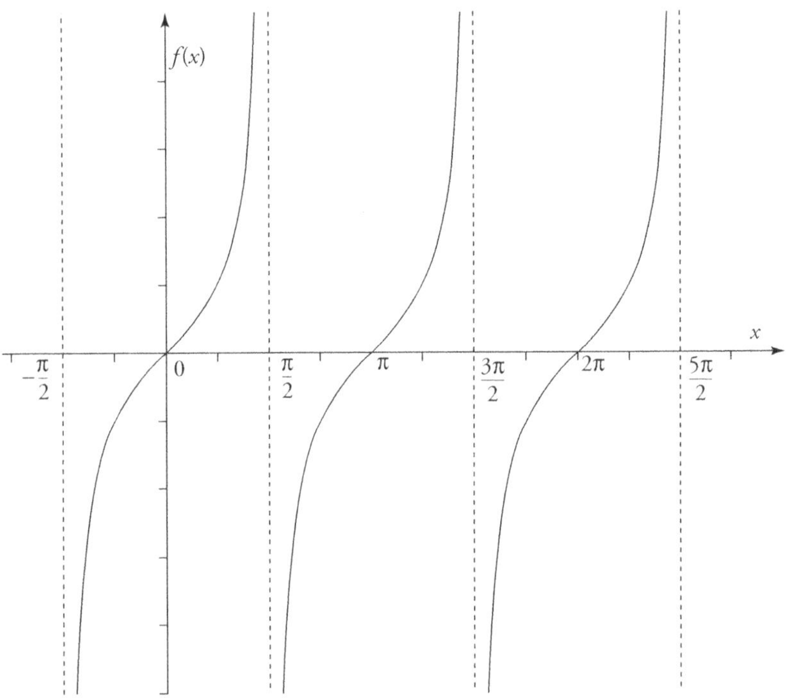

Since $\tan(-x) = \frac{\sin(-x)}{\cos(-x)} = \frac{-\sin x}{\cos x} = -\tan x$, the tangent function is an odd function. Verify that the graph of the tangent function is symmetrical with respect to the origin.

Cosecant Function

By definition, $\csc x = \frac{1}{\sin x}$. When $\sin x = 0$, the cosecant function is undefined. In the interval $[0, 2\pi]$, this happens at $x = 0$, $x = \pi$, and $x = 2\pi$. We can get a sketch of the cosecant function by looking at the graph of the sine function and using cosecant numbers of special angles and their multiples.

x	csc x	x	csc x
0	undefined	$\frac{7\pi}{6}$	–2
$\frac{\pi}{6}$	2	$\frac{5\pi}{4}$	$-\sqrt{2}$
$\frac{\pi}{4}$	$\sqrt{2}$	$\frac{4\pi}{3}$	$-\frac{2\sqrt{3}}{3}$
$\frac{\pi}{3}$	$\frac{2\sqrt{3}}{3}$	$\frac{3\pi}{2}$	–1
$\frac{\pi}{2}$	1	$\frac{5\pi}{3}$	$-\frac{2\sqrt{3}}{3}$
$\frac{2\pi}{3}$	$\frac{2\sqrt{3}}{3}$	$\frac{7\pi}{4}$	$-\sqrt{2}$
$\frac{3\pi}{4}$	$\sqrt{2}$	$\frac{11\pi}{6}$	–2
$\frac{5\pi}{6}$	2	2π	undefined
π	undefined		

As $x \to 0^+$, $\sin x \to 0^+$. That is, the denominator of $\frac{1}{\sin x}$ remains positive, the numerator is 1, and the denominator gets very small. Therefore, $\csc x \to +\infty$.

As $x \to \pi^-$, $\sin x \to 0^+$. That is, the denominator of $\frac{1}{\sin x}$ remains positive, the numerator is 1, and the denominator gets very small. Therefore, $\csc x \to +\infty$.

As $x \to \pi^+$, $\sin x \to 0^-$. That is, the denominator of $\frac{1}{\sin x}$ remains negative, the numerator is 1, and the denominator gets very small. Therefore, $\csc x \to -\infty$.

As $x \to 2\pi^-$, $\sin x \to 0^-$. That is, the denominator of $\frac{1}{\sin x}$ remains negative, the numerator is 1, and the denominator gets very small. Therefore, $\csc x \to -\infty$.

The cosecant function has asymptotes at $x = 0$, at $x = \pi$, and at $x = 2\pi$. The following is a sketch of the cosecant function in $(0, 2\pi)$.

There is no maximum value or minimum value. Therefore, there is no amplitude. The period is 2π.

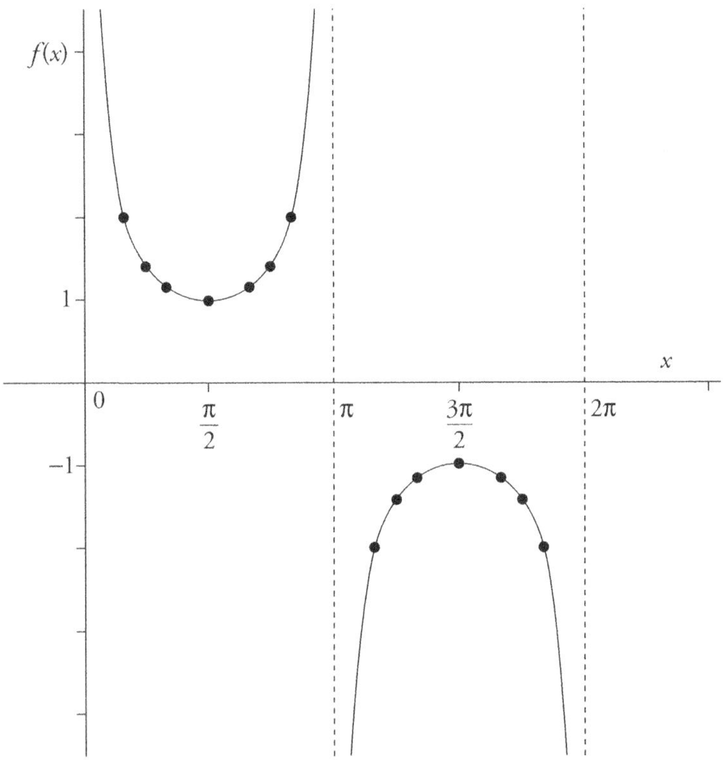

The cosecant function is an odd function. The following figure shows several cycles of the cosecant function.

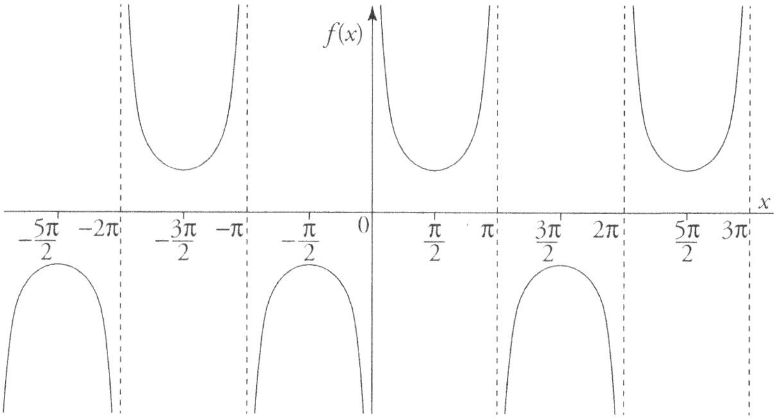

Secant Function

By definition, $\sec x = \dfrac{1}{\cos x}$. When $\cos x = 0$, the secant function is undefined. In the interval $[0, 2\pi]$, this happens at $x = \dfrac{\pi}{2}$ and at $x = \dfrac{3\pi}{2}$. We can get a sketch of the secant function by looking at the graph of the cosine function and using secant numbers of special angles and their multiples.

x	sec x	x	sec x
0	1	$\frac{7\pi}{6}$	$-\frac{2\sqrt{3}}{3}$
$\frac{\pi}{6}$	$\frac{2\sqrt{3}}{3}$	$\frac{5\pi}{4}$	$-\sqrt{2}$
$\frac{\pi}{4}$	$\sqrt{2}$	$\frac{4\pi}{3}$	–2
$\frac{\pi}{3}$	2	$\frac{3\pi}{2}$	undefined
$\frac{\pi}{2}$	undefined	$\frac{5\pi}{3}$	$\frac{2\sqrt{3}}{3}$
$\frac{2\pi}{3}$	$-\frac{2\sqrt{3}}{3}$	$\frac{7\pi}{4}$	$\sqrt{2}$
$\frac{3\pi}{4}$	$-\sqrt{2}$	$\frac{11\pi}{6}$	2
$\frac{5\pi}{6}$	–2	2π	1
π	–1		

As $x \to \frac{\pi^-}{2}$, $\cos x \to 0^+$. That is, the denominator of $\frac{1}{\cos x}$ remains positive, the numerator is 1, and the denominator gets very small. Therefore, $\sec x \to +\infty$.

As $x \to \frac{\pi^+}{2}$, $\cos x \to 0^-$. That is, the denominator of $\frac{1}{\cos x}$ remains negative, the numerator is 1, and the denominator gets very small. Therefore, $\sec x \to -\infty$.

As $x \to \frac{3\pi^-}{2}$, $\cos x \to 0^-$. That is, the denominator of $\frac{1}{\cos x}$ remains negative, the numerator is 1, and the denominator gets very small. Therefore, $\sec x \to -\infty$.

As $x \to \frac{3\pi^+}{2}$, $\cos x \to 0^+$. That is, the denominator of $\frac{1}{\cos x}$ remains positive, the numerator is 1, and the denominator gets very small. Therefore, $\sec x \to +\infty$.

The secant function has asymptotes at $x = \frac{\pi}{2}$ and at $x = \frac{3\pi}{2}$. The following is a sketch of the secant function in $[0, 2\pi]$.

There is no maximum value or minimum value. Therefore, there is no amplitude. The period is 2π.

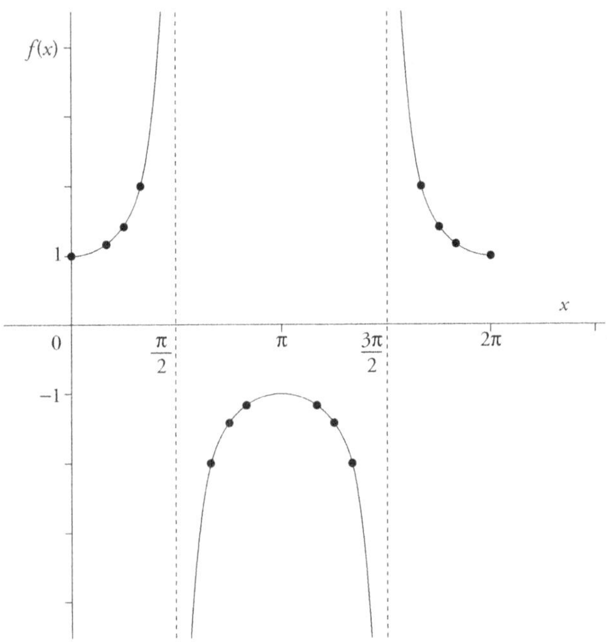

The secant function is an even function and the following figure shows several cycles of the secant function.

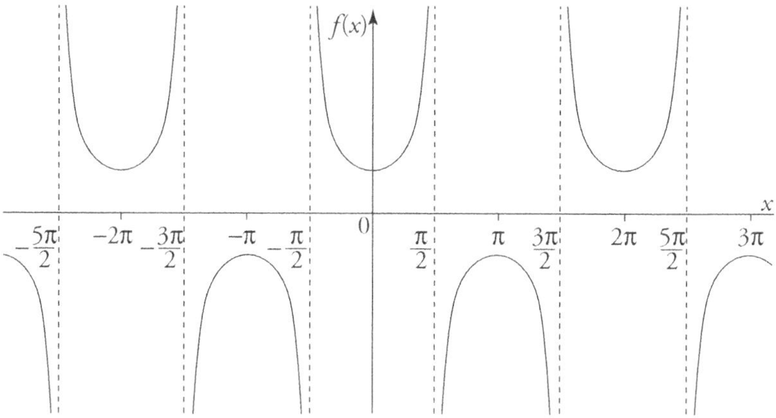

Cotangent Function

By definition $\cot x = \dfrac{1}{\tan x}$ and $\tan x = \dfrac{\sin x}{\cos x}$. Therefore, $\cot x = \dfrac{\cos x}{\sin x}$. We also know the period of the tangent function is π. When $\sin x = 0$, the cotangent function is undefined. In the interval $\left[-\dfrac{\pi}{2}, \dfrac{\pi}{2}\right]$, this happens at $x = 0$. We can get a sketch of the cotangent function by looking at the graphs of sine and cosine function together and using tangent numbers of special angles and their multiples in $\left[-\dfrac{\pi}{2}, \dfrac{\pi}{2}\right]$.

x	cot x	x	cot x
$-\dfrac{\pi}{2}$	0	$\dfrac{\pi}{6}$	$\sqrt{3}$
$-\dfrac{\pi}{6}$	$-\sqrt{3}$	$\dfrac{\pi}{4}$	1
$-\dfrac{\pi}{4}$	−1	$\dfrac{\pi}{3}$	$\dfrac{1}{\sqrt{3}}$
$-\dfrac{\pi}{3}$	$-\dfrac{1}{\sqrt{3}}$	$\dfrac{\pi}{2}$	0
0	undefined		

As $x \to 0^-$, $\cos x \approx 1$ and $\sin x \to 0^-$. That is, the numerator of $\frac{\cos x}{\sin x}$ remains close to 1, and the denominator is negative and gets very small. Therefore, $\cot x \to -\infty$.

As $x \to 0^+$, $\cos x \approx 1$ and $\sin x \to 0^+$. That is, the numerator of $\frac{\cos x}{\sin x}$ remains close to 1, and the denominator is positive and gets very small. Therefore, $\cot x \to +\infty$.

The cotangent function has an asymptote at $x = 0$. The following figure shows the graph of the cotangent function in the interval $\left[-\frac{\pi}{2}, \frac{\pi}{2}\right]$.

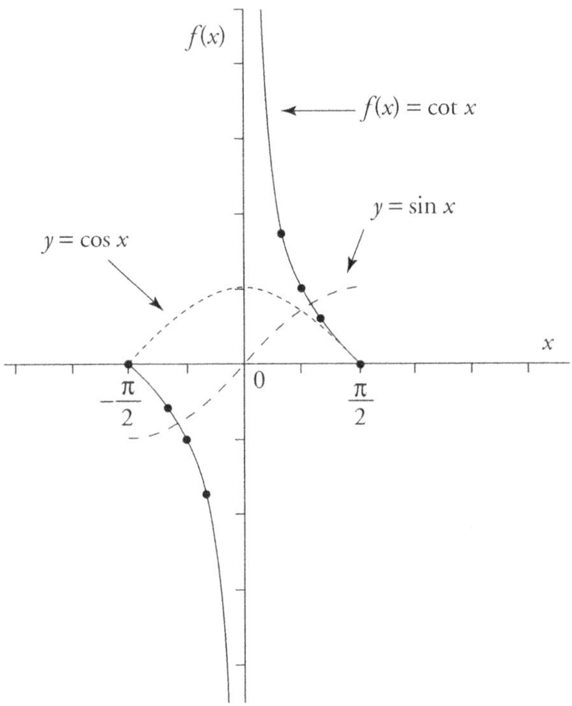

There is no amplitude, and the period is π. The cotangent function is an odd function. The following figure shows several cycles of the cotangent function.

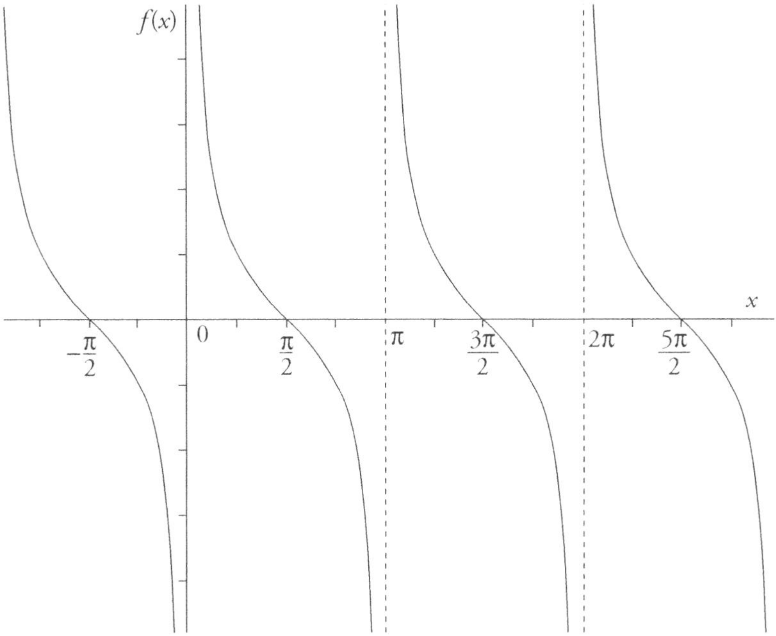

Chapter Check-Out

Questions

1. What is the period of the sine function?
2. What is the period of the cosine function?
3. What is the period of the tangent function?
4. What is the amplitude of the sine function?
5. True or False: The graphs of both the sine function and the tangent function lie below the x-axis in the fourth quadrant.
6. True or False: The sine function and the tangent function are odd functions.
7. True or False: The cosine function is an even function.
8. What is the period of the cotangent function?

Answers

1. 2π
2. 2π
3. π

4. 1
5. True
6. True
7. True
8. π

Chapter 4

INVERSE TRIGONOMETRIC FUNCTIONS

Chapter Check-In

- ❑ Defining inverse trigonometric functions
- ❑ Demonstrating how to restrict the basic trigonometric functions to certain quadrants
- ❑ Showing that the restricted trigonometric functions are one-to-one and have inverses
- ❑ Solving problems using inverse trigonometric functions

Common Core Standard: Model periodic phenomena with trigonometric functions

Understand that restricting a trigonometric function to a domain on which it is always increasing or always decreasing allows its inverse to be constructed. (HSF.TF.B.6)

Use inverse functions to solve trigonometric equations that arise in modeling contexts; evaluate the solutions using technology and interpret them in terms of the context. (HSF.TF.B.7)

Please also refer to the standards for mathematical practice, found here: www.corestandards.org/Math/Practice/.

The standard trigonometric functions are periodic, meaning that they repeat themselves. Therefore, the same output value appears for multiple input values of the function. This makes inverse functions of the standard trigonometric functions impossible to construct. In order to solve equations involving trig functions, it is imperative for inverse functions to exist. Thus, mathematicians have to restrict the trig functions in order create these inverses.

Trigonometry Theorems

The trigonometry theorems in this book are numbered for organizational purposes. For example, *Theorem 2.2* is the second theorem given in Chapter 2. Study these theorems by their content, not by their number, as the theorem numbering has no significance outside of this book.

Inverse Functions

To define an inverse function, the original function must be **one-to-one (1-1).** For a one-to-one correspondence to exist, (1) each value in the domain must correspond to exactly one value in the range, and (2) each value in the range must correspond to exactly one value in the domain. The first restriction is shared by all functions; the second is not. Just like the vertical line test that can be used to test whether a given graph is a graph of a function, a horizontal line test can be used to check if a given function is 1-1. If a horizontal line passing through any given point of the range of the function intersects the graph of the function at only one point, then for that point in the range, there is exactly one corresponding value in the domain. The sine function, for example, does not satisfy the second restriction, since the same value in the range corresponds to many values in the domain (by the horizontal line test), as seen in the following figure.

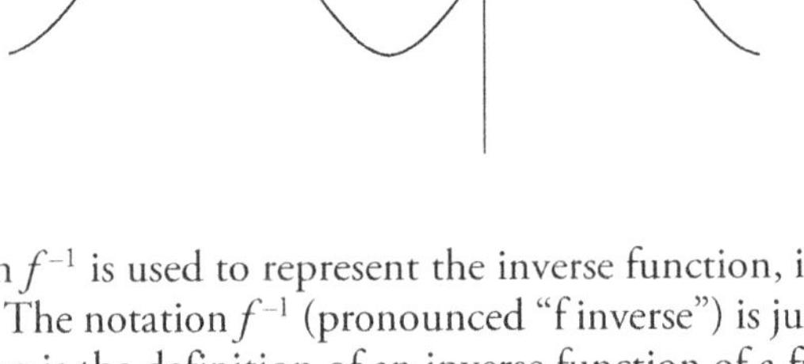

The notation f^{-1} is used to represent the inverse function, if it exists, for a function f. The notation f^{-1} (pronounced "f inverse") is just a notation. The following is the definition of an inverse function of a function.

Suppose a function f is 1-1. Then f^{-1} exists and

(a) $f^{-1}(f(x)) = x$, for any x in the domain of f, and

(b) $f(f^{-1}(x)) = x$, for any x in the domain of f^{-1}.

Since f is 1-1, there is one inverse function for f and vice versa.

This definition implies that the inverse function "undoes" what a function does and vice versa. Part (a) of the 1-1 definition states that for any point x in the domain of f, the function f carries it to $f(x)$ and the inverse function carries $f(x)$ back to x. Part (b) of the definition states that for any point x in the domain of f^{-1}, the function f^{-1} carries it to $f^{-1}(x)$ and the function carries $f^{-1}(x)$ back to x.

Theorem 4.1: Let $P(a, b)$ be a point on the xy-coordinate plane. Then the reflection of P across the line $y = x$ is the point (b, a).

Proof:

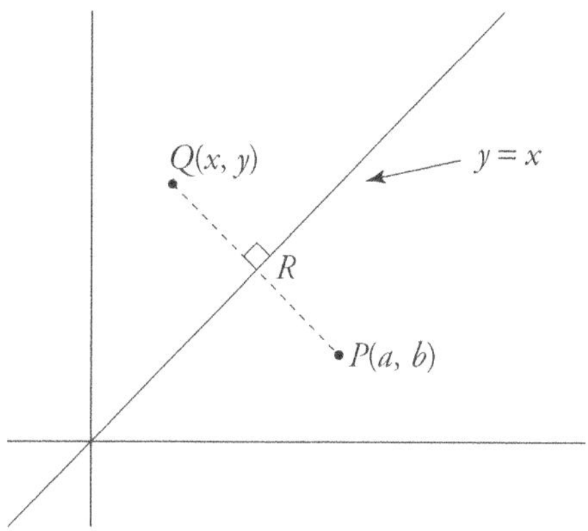

Let $Q(x, y)$ be the reflection of P across the line $y = x$. Let R be the point of intersection of the segment $\overline{PQ}$ and the line $y = x$. Then the line $\overleftrightarrow{PQ}$ is perpendicular to the line $y = x$ and $PR = RQ$. The line $\overleftrightarrow{PQ}$ has the slope $\frac{y-b}{x-a}$ and the line $y = x$ has the slope 1. Since line $\overleftrightarrow{PQ}$ is perpendicular to the line $y = x$, the product of their slopes is equal to -1. Therefore,

$$\frac{y-b}{x-a} = -1$$

$$y - b = -x + a \qquad (1)$$

Since R is the midpoint of the segment PQ, R has coordinates $\left(\frac{x+a}{2}, \frac{y+b}{2}\right)$. Since R is on the line $y = x$,

$$\frac{x+a}{2} = \frac{y+b}{2}$$
$$y+b = x+a \qquad (2)$$

By adding equations (1) and (2), you get $y = a$, and by subtracting equation (1) from equation (2), you get $x = b$.

The graph of the line $y = x$ is known as the **diagonal.**

Theorem 4.2: Suppose f is a 1-1 function. Then the graph of the inverse function is the reflection of the graph of f across the diagonal.

Proof: Let g be the function whose graph is obtained by reflecting the graph of f across the diagonal. Let $(x, f(x))$ be an arbitrary point on the graph of f. Then by *Theorem 4.1*, $(f(x), x)$ is a point on the graph of g. That is, $g(f(x)) = x$. But, $f^{-1}(f(x)) = x$. Since f^{-1} is unique, $g = f^{-1}$. Since x is arbitrary, the reflection of the graph of f across the diagonal is the graph of f^{-1}.

Inverse Sine Function

We know that $-1 \leq \sin x \leq 1$, and all possible values of $\sin x$ lie in the interval $\left[-\frac{\pi}{2}, \frac{\pi}{2}\right]$. If we restrict the domain of the sine function to the interval $\left[-\frac{\pi}{2}, \frac{\pi}{2}\right]$, not only do we have all possible sine function values in the interval, but it also makes the restricted sine function 1-1.

We define the inverse sine function, denoted by $\sin^{-1} x$, as follows:

$\sin^{-1}(\sin x) = x$, for all x in $\left[-\frac{\pi}{2}, \frac{\pi}{2}\right]$, and

$\sin(\sin^{-1} x) = x$, for all x in $[-1, 1]$

By *Theorem 4.2*, the reflection of the restricted sine function across the diagonal is the graph of the inverse sine function.

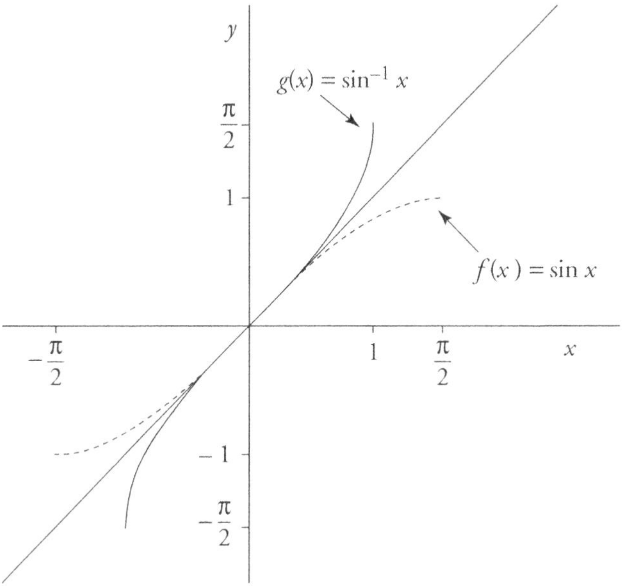

The domain of the inverse sine function is $[-1, 1]$. That is, $\sin^{-1} x$ makes sense if $-1 \le x \le 1$. The range of the inverse sine function is $\left[-\frac{\pi}{2}, \frac{\pi}{2}\right]$. That is, $-\frac{\pi}{2} \le \sin^{-1} x \le \frac{\pi}{2}$.

Inverse Cosine Function

We know that $-1 \le \cos x \le 1$, and all possible values of $\cos x$ lie in the interval $[0, \pi]$. If we restrict the domain of the cosine function to the interval $[0, \pi]$, then all the cosine function values are in the interval, and the restricted cosine function is 1-1.

We define the inverse cosine function, denoted by $\cos^{-1} x$, as follows:

$\cos^{-1}(\cos x) = x$, for all x in $[0, \pi]$, and

$\cos(\cos^{-1} x) = x$, for all x in $[-1, 1]$

By *Theorem 4.2*, the reflection of the restricted cosine function across the diagonal is the graph of the inverse cosine function.

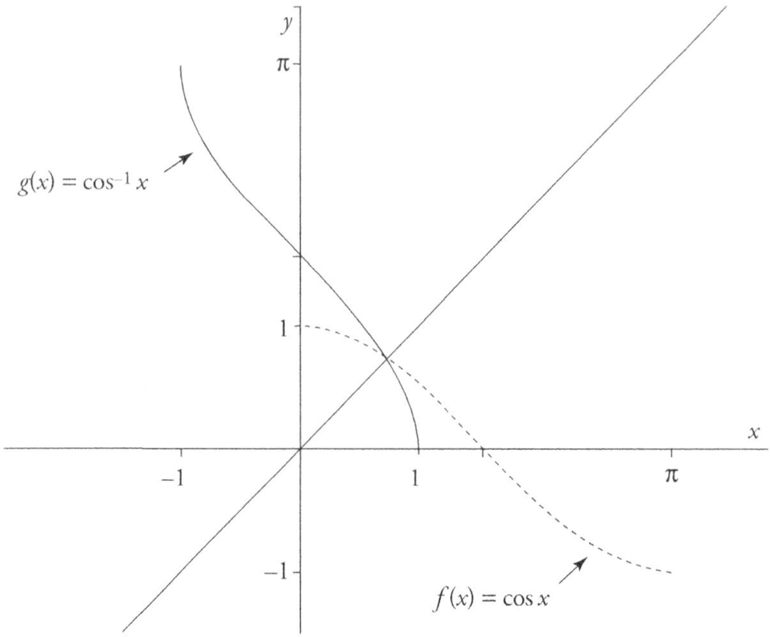

The domain of the inverse cosine function is $[-1, 1]$. That is, $\cos^{-1} x$ makes sense if $-1 \leq x \leq 1$. The range of the inverse cosine function is $[0, \pi]$. That is, $0 \leq \cos^{-1} x \leq \pi$.

Inverse Tangent Function

All possible values of $\tan x$ lie in the interval $\left(-\frac{\pi}{2}, \frac{\pi}{2}\right)$. If we restrict the domain of the tangent function to the interval $\left(-\frac{\pi}{2}, \frac{\pi}{2}\right)$, then we have all possible tangent function values in the interval, and the restricted tangent function is 1-1.

We define the inverse tangent function, denoted by $\tan^{-1} x$, as follows:

$\tan^{-1}(\tan x) = x$, for all x in $\left(-\frac{\pi}{2}, \frac{\pi}{2}\right)$, and

$\tan(\tan^{-1} x) = x$, for all x

By *Theorem 4.2*, the reflection of the restricted tangent function across the diagonal is the graph of the inverse tangent function. The domain of the inverse tangent function is $(-\infty, \infty)$. That is, $\tan^{-1} x$ makes sense for all real numbers x. The range of the inverse tangent function is $\left(-\frac{\pi}{2}, \frac{\pi}{2}\right)$. That is, $-\frac{\pi}{2} < \tan^{-1} x < \frac{\pi}{2}$.

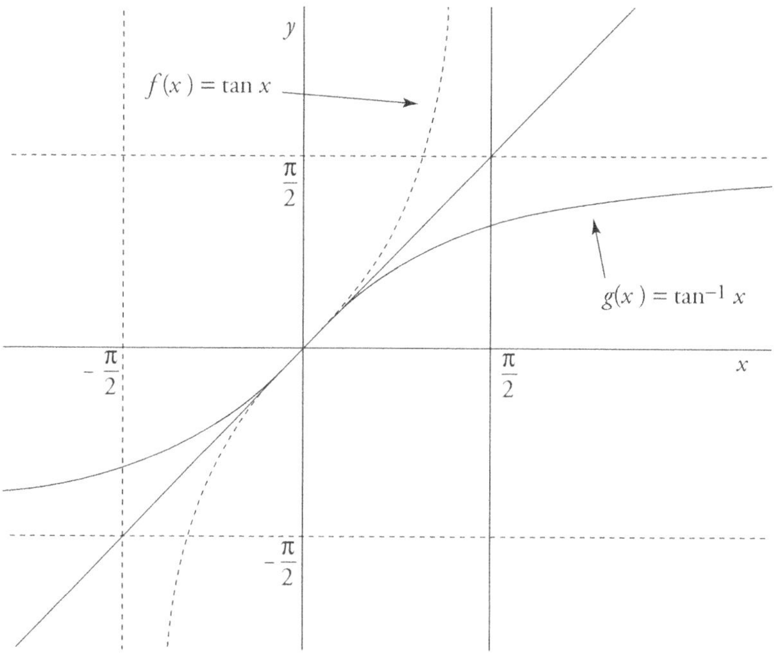

Example 1: Find the exact value of $\cos^{-1}\left(-\frac{\sqrt{3}}{2}\right)$.

Let $x = \cos^{-1}\left(-\frac{\sqrt{3}}{2}\right)$. Then $0 \le x \le \pi$.

$$\cos x = \cos\left(\cos^{-1}\left(-\frac{\sqrt{3}}{2}\right)\right)$$

$$\cos x = -\frac{\sqrt{3}}{2}$$

by the definition of the cosine inverse function.

Since $\cos x < 0$, this particular x lies in the second quadrant. This is the known angle $x = \frac{5\pi}{6}$. Therefore, $\cos^{-1}\left(-\frac{\sqrt{3}}{2}\right) = \frac{5\pi}{6}$.

Example 2: Find the exact value of $\sin^{-1}\left(\frac{\sqrt{2}}{2}\right)$.

Let $x = \sin^{-1}\left(\frac{\sqrt{2}}{2}\right)$. Then $\sin x = \frac{\sqrt{2}}{2}$ and $-\frac{\pi}{2} \le x \le \frac{\pi}{2}$. Since $\sin x > 0$, this particular x is in the first quadrant. This is the special angle $\frac{\pi}{4}$. Therefore, $\sin^{-1}\left(\frac{\sqrt{2}}{2}\right) = \frac{\pi}{4}$.

Example 3: Find the exact value of $\cos(\cos^{-1} 0.62)$, if possible.

Since $-1 \le 0.62 \le 1$, by the definition of cosine inverse function, $\cos(\cos^{-1} 0.62) = 0.62$.

Example 4: Find the exact value of $\sin(\sin^{-1} 1.02)$, if possible.

The number 1.02 is not in the interval $[-1, 1]$, and there is no $\sin^{-1} 1.02$. Therefore, $\sin(\sin^{-1} 1.02)$ does not exist.

Example 5: Find the exact value of $\sin^{-1}\left(\sin\frac{2\pi}{3}\right)$, if possible.

The number $\frac{2\pi}{3}$ is not in the interval $\left[-\frac{\pi}{2}, \frac{\pi}{2}\right]$. That means $\sin^{-1}\left(\sin\frac{2\pi}{3}\right) \ne \frac{2\pi}{3}$. However, $\sin\frac{2\pi}{3}$ is a known number, and by using the unit circle, you can easily verify that $\sin\frac{2\pi}{3} = \sin\frac{\pi}{3}$.

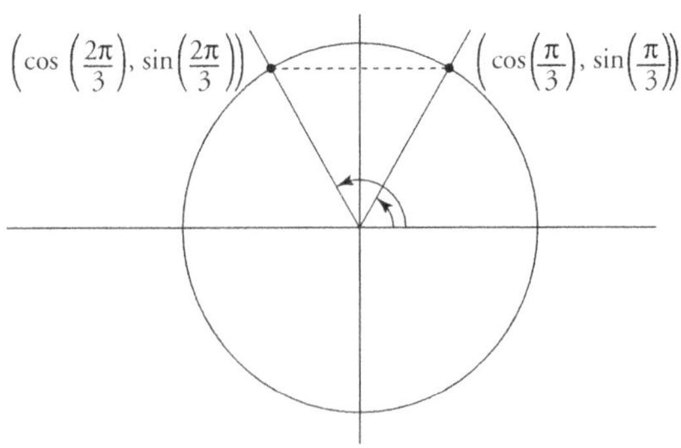

Therefore, $\sin^{-1}\left(\sin\frac{2\pi}{3}\right) = \sin^{-1}\left(\sin\frac{\pi}{3}\right) = \frac{\pi}{3}$, since $\frac{\pi}{3}$ is in the interval $\left[-\frac{\pi}{2}, \frac{\pi}{2}\right]$.

Example 6: Find the exact value of $\cos^{-1}\left(\cos\frac{7\pi}{6}\right)$, if possible.

The number $\frac{7\pi}{6}$ is not in the interval $[0, \pi]$. That means $\cos^{-1}\left(\cos\frac{7\pi}{6}\right) \neq \frac{7\pi}{6}$. However, $\cos\frac{7\pi}{6}$ is a known number, and by using the unit circle, you can easily verify that $\cos\frac{7\pi}{6} = \cos\frac{5\pi}{6}$.

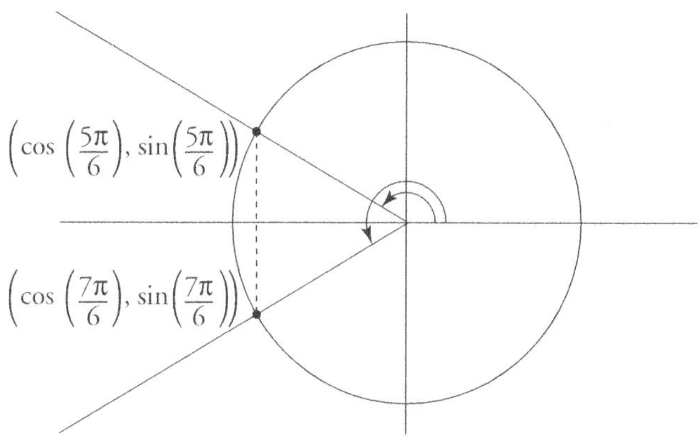

Therefore, $\cos^{-1}\left(\cos\frac{7\pi}{6}\right) = \cos^{-1}\left(\cos\frac{5\pi}{6}\right) = \frac{5\pi}{6}$, since $\frac{5\pi}{6}$ is in the interval $[0, \pi]$.

Example 7: Find the exact value of $\sin^{-1}\left(\sin\frac{4\pi}{3}\right)$, if possible.

The number $\frac{4\pi}{3}$ is not in the interval $\left[-\frac{\pi}{2}, \frac{\pi}{2}\right]$. That means $\sin^{-1}\left(\sin\frac{4\pi}{3}\right) \neq \frac{4\pi}{3}$. However, $\sin\frac{4\pi}{3}$ is a known number, and by using the unit circle, you can easily verify that $\sin\frac{4\pi}{3} = \sin\left(-\frac{\pi}{3}\right)$.

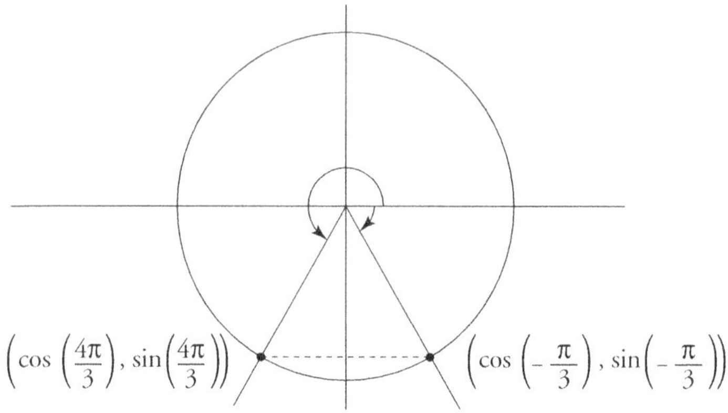

Therefore, $\sin^{-1}\left(\sin\frac{4\pi}{3}\right) = \sin^{-1}\left(\sin\left(-\frac{\pi}{3}\right)\right) = -\frac{\pi}{3}$, since $-\frac{\pi}{3}$ is in the interval $\left[-\frac{\pi}{2}, \frac{\pi}{2}\right]$.

Example 8: Find the exact value of $\tan^{-1}\left(\tan\frac{4\pi}{3}\right)$, if possible.

The number $\frac{4\pi}{3}$ is not in the interval $\left[-\frac{\pi}{2}, \frac{\pi}{2}\right]$. That means $\tan^{-1}\left(\tan\frac{4\pi}{3}\right) \neq \frac{4\pi}{3}$. However, $\tan\frac{4\pi}{3}$ is a known number, and by using the unit circle, you can verify that $\tan\frac{4\pi}{3} = \tan\left(\frac{\pi}{3}\right)$.

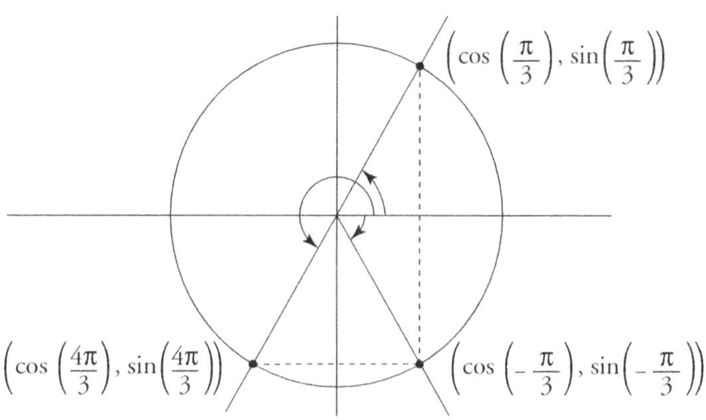

Therefore, $\tan^{-1}\left(\tan\frac{4\pi}{3}\right)=\tan^{-1}\left(\tan\left(\frac{\pi}{3}\right)\right)=\frac{\pi}{3}$, since $\frac{\pi}{3}$ is in the interval $\left[-\frac{\pi}{2},\frac{\pi}{2}\right]$. The relationship $\tan\frac{4\pi}{3}=\tan\left(\frac{\pi}{3}\right)$ can be seen more clearly, perhaps, using the graph of the tangent function.

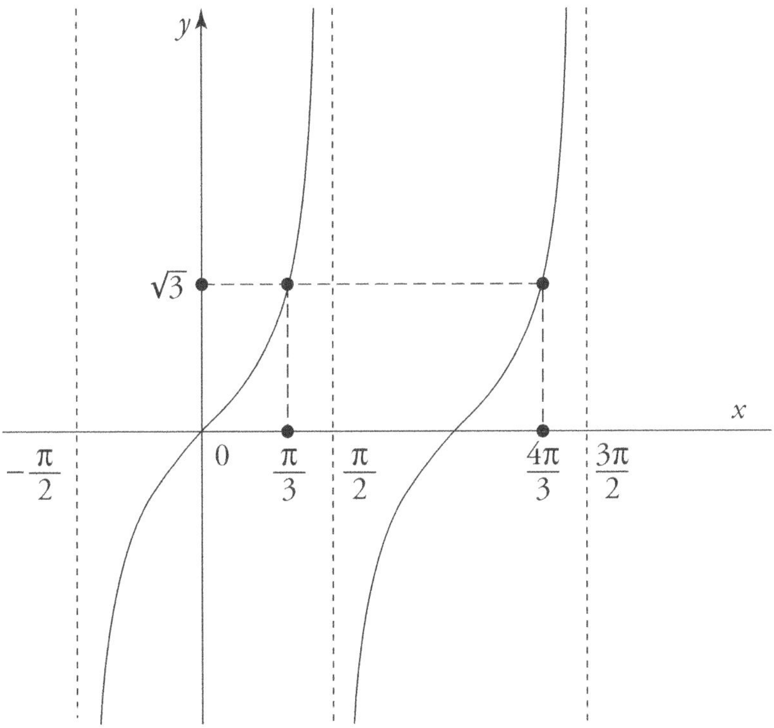

Example 9: Find the exact value of $\sin\left(\cos^{-1}\left(\frac{\sqrt{3}}{2}\right)\right)$, if possible.

Let $x=\cos^{-1}\left(\frac{\sqrt{3}}{2}\right)$. Then $\cos x=\frac{\sqrt{3}}{2}$, where $0\le x\le\pi$. However, since $\cos x>0$, x is in the first quadrant, or $0<x<\frac{\pi}{2}$.

Since $\sin^2 x + \cos^2 x = 1$ for any x,

$$\sin^2 x + \left(\frac{\sqrt{3}}{2}\right)^2 = 1$$

$$\sin^2 x + \frac{3}{4} = 1$$

$$\sin^2 x = 1 - \frac{3}{4}$$

$$\sin^2 x = \frac{1}{4}$$

$$\sin x = \pm\frac{1}{2}$$

Since x is in the first quadrant, $\sin x = \frac{1}{2}$.

Example 10: Find the exact value of $\cos\left(\sin^{-1}\left(-\frac{\sqrt{3}}{2}\right)\right)$, if possible.

Let $x = \sin^{-1}\left(-\frac{\sqrt{3}}{2}\right)$. Then $\sin x = -\frac{\sqrt{3}}{2}$, where $-\frac{\pi}{2} \le x \le \frac{\pi}{2}$. However, since $\sin x < 0$, x is in the fourth quadrant, or $-\frac{\pi}{2} \le x \le 0$.
Since $\cos^2 x + \sin^2 x = 1$ for any x,

$$\cos^2 x + \left(-\frac{\sqrt{3}}{2}\right)^2 = 1$$

$$\cos^2 x = 1 - \frac{3}{4}$$

$$\cos^2 x = \frac{1}{4}$$

$$\cos x = \pm\frac{1}{2}$$

Since x is in the fourth quadrant, $\cos x = \frac{1}{2}$.

Other Inverse Trigonometric Functions

The remaining three trigonometric functions are not 1-1 either. We will restrict their domains as we did with sine, cosine, and tangent functions. We define the inverse cosecant function as follows:

$$\csc^{-1}(\csc x) = x, \text{ for all } x \text{ in } \left[-\frac{\pi}{2}, 0\right) \cup \left(0, \frac{\pi}{2}\right], \text{ and}$$

$$\csc(\csc^{-1} x) = x, \text{ for all } x \text{ in } (-\infty, -1] \cup [1, \infty)$$

The following figure shows the graphs of both the restricted cosecant function and the inverse cosecant function.

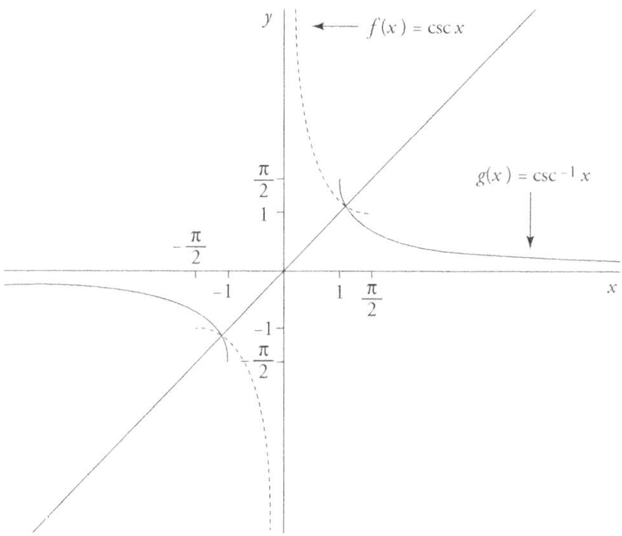

The domain of the inverse cosecant function is $(-\infty,-1] \cup [1,\infty)$, and the range of the inverse cosecant function is $\left[-\frac{\pi}{2},0\right) \cup \left(0,\frac{\pi}{2}\right]$. That is, $-\frac{\pi}{2} \leq \csc^{-1} x < 0$ or $0 < \csc^{-1} x \leq \frac{\pi}{2}$.

We define the inverse secant function as follows:

$$\sec^{-1}(\sec x) = x, \text{ for all } x \text{ in } \left[0, \frac{\pi}{2}\right) \cup \left(\frac{\pi}{2}, \pi\right], \text{ and}$$

$$\sec(\sec^{-1} x) = x, \text{ for all } x \text{ in } (-\infty, -1] \cup [1, \infty)$$

The following figure shows the graphs of both the restricted secant function and the inverse secant function.

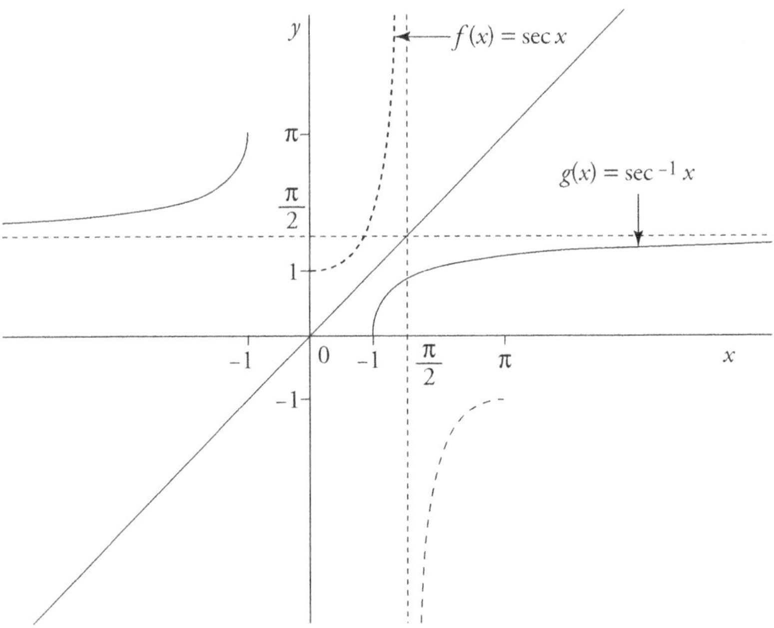

The domain of the inverse secant function is $(-\infty, -1] \cup [1, \infty)$, and the range of the inverse secant function is $\left[0, \frac{\pi}{2}\right) \cup \left(\frac{\pi}{2}, \pi\right]$. That is, $0 \le \sec^{-1} x < \frac{\pi}{2}$ or $\frac{\pi}{2} < \sec^{-1} x \le \pi$.

We define the inverse cotangent function as follows:

The restricted cotangent function has a hole at $x = -\frac{\pi}{2}$ to make it a 1-1 function.

$\cot^{-1}(\cot x) = x$, for all x in $\left(-\frac{\pi}{2}, 0\right) \cup \left(0, \frac{\pi}{2}\right]$, and

$\cot\left(\cot^{-1} x\right) = x$, for all x in $(-\infty, \infty)$

The following figure shows the graphs of both the restricted cotangent function and the inverse cotangent function.

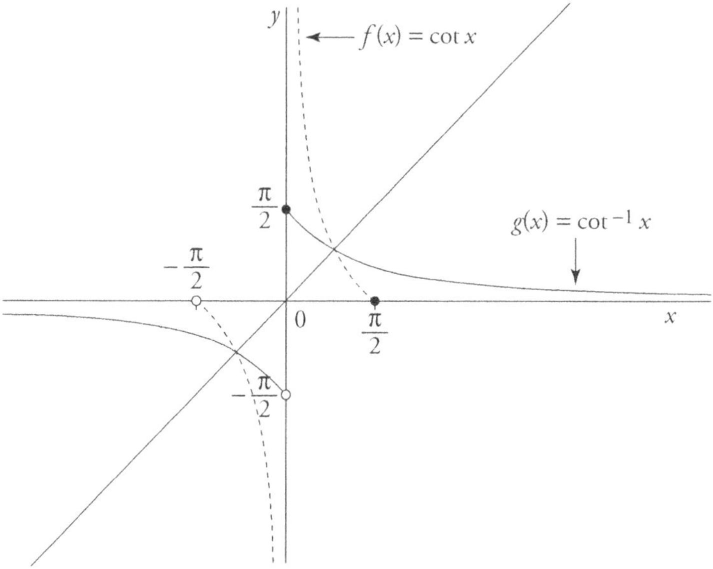

The domain of the inverse cotangent function is $(-\infty, \infty)$, and the range of the inverse cotangent function is $\left(-\frac{\pi}{2}, 0\right) \cup \left(0, \frac{\pi}{2}\right]$. That is, $-\frac{\pi}{2} < \cot^{-1} x < 0$ or $0 < \cot^{-1} x \leq \frac{\pi}{2}$.

Example 11: Determine the exact value of $\sin(\sec^{-1}(-4))$ without using a calculator or tables of trigonometric functions.

If $x = \sec^{-1}(-4)$, then $\sec x = -4$ and $\cos x = -\frac{1}{4}$, where $0 \leq x \leq \pi$, $x \neq \frac{\pi}{2}$.

However, $\cos x < 0$ implies that x is in the second quadrant. Since $\sin^2 x + \cos^2 x = 1$ for any x,

$$\sin^2 x + \left(-\frac{1}{4}\right)^2 = 1$$

$$\sin^2 x = 1 - \frac{1}{16}$$

$$\sin^2 x = \frac{15}{16}$$

$$\sin x = \pm\frac{\sqrt{15}}{4}$$

Since x is in the second quadrant, $\sin x = \frac{\sqrt{15}}{4}$.

Example 12: Determine the exact value of cos ($\tan^{-1}$ (7)) without using a calculator or tables of trigonometric functions.

Let $x = \tan^{-1}(7)$. Then $\tan x = 7$, where $-\frac{\pi}{2} < x < \frac{\pi}{2}$. Since $\tan x > 0$, x is in the first quadrant. Therefore, both $\sin x$ and $\cos x$ are positive. Since $\sin^2 x + \cos^2 x = 1$,

$$\frac{\sin^2 x}{\cos^2 x} + \frac{\cos^2 x}{\cos^2 x} = \frac{1}{\cos^2 x}$$

$$\tan^2 x + 1 = \sec^2 x$$

$$7^2 + 1 = \sec^2 x$$

$$\sec^2 x = 50$$

$$\sec x = \pm 5\sqrt{2}$$

Since x is in the first quadrant, $\sec x = 5\sqrt{2}$ and $\cos x = \frac{1}{5\sqrt{2}}$.

Chapter Check-Out

Questions

1. Find the exact value of $\cos\left(\sin^{-1}\left(-\frac{3}{7}\right)\right)$.

2. Find the exact value of $\tan\left(\sin^{-1}\left(\frac{2}{3}\right)\right)$.

3. True or False: The inverse sine and inverse cosine are defined in the same quadrants.

4. Find the exact value of $\sin\left(\sin^{-1}\left(\frac{3}{8}\right)\right)$.

Answers

1. $\frac{2\sqrt{10}}{7}$

2. $\frac{2\sqrt{5}}{5}$

3. False

4. $\frac{3}{8}$

Chapter 5

BASIC TRIGONOMETRIC EQUATIONS

Chapter Check-In

- ❑ Defining basic trigonometric equations
- ❑ Demonstrating how to find the general solution of a basic trigonometric equation
- ❑ Demonstrating how to find the solutions of a basic trigonometric equation in an interval

Common Core Standard: Model periodic phenomena with trigonometric functions

Use inverse functions to solve trigonometric equations that arise in modeling contexts; evaluate the solutions using technology, and interpret them in terms of the context. (HSF.TF.B.7)

Please also refer to the standards for mathematical practice, found here: www.corestandards.org/Math/Practice/.

The equations of the type $\sin x = a$, $\cos x = a$, and $\tan x = a$, where a is a real number, are called basic trigonometric equations. These equations make no sense for some real numbers a. For example, the equation $\sin x = 2$ makes no sense because for all x, $\sin x$ is in the interval $[-1, 1]$. Such an equation has no solutions. Therefore, we will put restrictions on the number a whenever necessary.

Trigonometry Theorems

The trigonometry theorems in this book are numbered for organizational purposes. For example, *Theorem 2.2* is the second theorem given in Chapter 2. Study these theorems by their content, not by their number, as the theorem numbering has no significance outside of this book.

Basic Sine Equation

The equation $\sin\theta = a$, where $-1 \le a \le 1$, is called the *basic sine equation.* Our goal is to find *all* solutions of the basic sine equation. We will use the unit circle to find these solutions.

Without a loss of generality, we will choose $a > 0$. Consider the unit circle and let the points of intersection of the unit circle and the line $y = a$ be B and C, as shown in the following figure. Let O be the origin, let A be the point of intersection of the unit circle and the positive x-axis, and let D be the point of intersection of the unit circle and the negative x-axis. All angle measures are in radians for convenience.

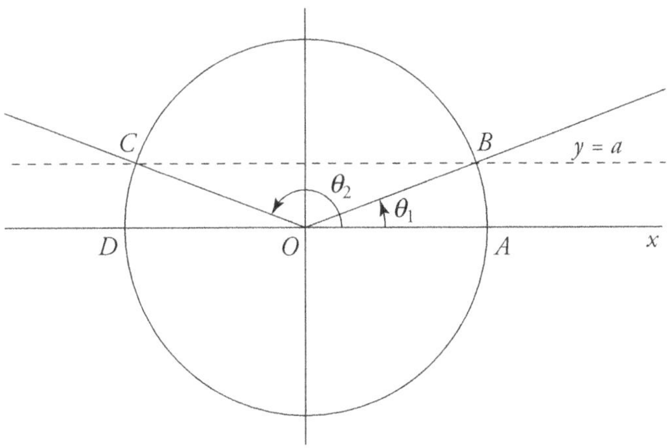

Let $\angle AOB = \theta_1$ and $\angle AOC = \theta_2$. Then, by *Theorem 2.6*, $\sin\theta_1 = a$ and $\sin\theta_2 = a$. Therefore, θ_1 and θ_2 are solutions of the equation $\sin\theta = a$. The lines $y = a$ and $y = 0$ (the x-axis) are parallel, and line $\overleftrightarrow{OB}$ is a transversal; therefore, $\angle OBC = \angle AOB = \theta_1$. ΔOBC is an isosceles triangle with sides $OB = OC$. Therefore, $\angle OCB = \angle OBC = \theta_1$. The lines $y = a$ and $y = 0$ are parallel, and line $\overleftrightarrow{OC}$ is a transversal; therefore, $\angle DOC = \angle OCB = \theta_1$. Therefore, $\theta_2 = \pi - \theta_1$.

As a result, the following counterclockwise angles in standard position are solutions of the equation $\sin\theta = a$:

$$\theta_1,\ \pi - \theta_1,\ 2\pi + \theta_1,\ 3\pi - \theta_1,\ 4\pi + \theta_1,\ 5\pi - \theta_1,\ \ldots$$

Also, the following clockwise angles in standard position are solutions of the equation $\sin\ \theta = a$:

$$-\pi - \theta_1, -2\pi + \theta_1, -3\pi - \theta_1, -4\pi + \theta_1, -5\pi - \theta_1, \ldots$$

This infinite collection of angles are the solutions of the equation $\sin\ \theta = a$. We can write all these solutions in a condensed form if we observe certain patterns of these solutions. First, notice that each solution is a sum of a multiple of π and θ_1. Second, the signs of θ_1 in any two consecutive solutions are opposite. The number $(-1)^n$, where n is an integer, provides us with a convenient sign changer because $(-1)^n = 1$ for even n and $(-1)^n = -1$ for odd n. The following statement contains all solutions of the equation $\sin\ \theta = a$:

$$\theta = n\pi + (-1)^n\theta_1, \text{ where } n \text{ is an integer.}$$

We can do a little better. Since θ_1 is a solution of $\sin\ \theta = a$, we have

$$\sin\theta_1 = a.$$

$$\text{Then, } \sin^{-1}\left(\sin\theta_1\right) = \sin^{-1} a.$$

$$\text{Since } -\frac{\pi}{2} \le \theta_1 \le \frac{\pi}{2},\ \sin^{-1}\left(\sin\theta_1\right) = \theta_1.$$

$$\text{Therefore, } \theta_1 = \sin^{-1} a.$$

Therefore, the solutions of the equation $\sin\ \theta = a$ are $n\pi + (-1)^n \sin^{-1}a$, where n is an integer. This collection of solutions is known as the **general solution** of the equation $\sin\ \theta = a$.

Theorem 5.1: The general solution of $\sin\ \theta = a$ is $n\pi + (-1)^n \sin^{-1}a$, where n is an integer and $-1 \le a \le 1$.

Example 1: Find the general solution of the equation $\sin\theta = \frac{1}{2}$, and find solutions of the equation in the interval $[0, \pi]$.

By *Theorem 5.1*, the general solution is $n\pi + (-1)^n \sin^{-1}\left(\frac{1}{2}\right)$, where n is an integer. Since $\sin^{-1}\left(\frac{1}{2}\right) = \frac{\pi}{6}$, the general solution is $n\pi + (-1)^n \frac{\pi}{6}$.

When $n = 0$, we get $(-1)^0 \frac{\pi}{6} = \frac{\pi}{6}$, and this solution is in $[0, \pi]$.

When $n = 1$, we get $\pi + (-1)^1 \frac{\pi}{6} = \pi - \frac{\pi}{6} = \frac{5\pi}{6}$, and this solution is also in $[0, \pi]$.

When $n = 2$, we get $2\pi + (-1)^2 \frac{\pi}{6} = 2\pi + \frac{\pi}{6}$, and this solution is outside of $[0, \pi]$. It should be clear that for higher values of n, we will get solutions not in $[0, \pi]$.

When $n = -1$, we get $-\pi + (-1)^{-1} \frac{\pi}{6} = -\pi - \frac{\pi}{6}$, and this solution is not in $[0, \pi]$. It should be clear that for lower values of n, we will get solutions not in $[0, \pi]$.

Therefore, there are two solutions of the equation $\sin\theta = \frac{1}{2}$ in the interval $[0, \pi]$: $\frac{\pi}{6}$ and $\frac{5\pi}{6}$.

Example 2: Find the general solution of the equation $\sin\theta = -\frac{\sqrt{3}}{2}$, and find solutions in the interval $[0, \pi]$.

By *Theorem 5.1*, the general solution is $n\pi + (-1)^n \sin^{-1}\left(-\frac{\sqrt{3}}{2}\right)$, where n is an integer. Since $\sin^{-1}\left(-\frac{\sqrt{3}}{2}\right) = -\frac{\pi}{3}$, the general solution is $n\pi - (-1)^n \frac{\pi}{3}$.

When $n = 0$, we get $-(-1)^0 \frac{\pi}{3} = -\frac{\pi}{3}$, and this solution is not in $[0, \pi]$.

When $n = 1$, we get $\pi - (-1)^1 \frac{\pi}{3} = \pi + \frac{\pi}{3}$, and this solution is also not in $[0, \pi]$.

When $n = 2$, we get $2\pi - (-1)^2 \frac{\pi}{3} = 2\pi - \frac{\pi}{3}$, and this solution is not in $[0, \pi]$. It should be clear that for higher values of n, we will get solutions not in $[0, \pi]$.

When $n = -1$, we get $-\pi - (-1)^{-1} \frac{\pi}{3} = -\pi + \frac{\pi}{3} = -\frac{2\pi}{3}$, and this solution is not in $[0, \pi]$. It should be clear that for lower values of n, we will get solutions not in $[0, \pi]$. Therefore, there are no solutions of the equation $\sin\theta = -\frac{\sqrt{3}}{2}$ in the interval $[0, \pi]$.

Example 3: Find the general solution of the equation $\sin \theta = 0$, and find solutions in the interval $[0, \pi]$.

By *Theorem 5.1*, the general solution is $n\pi + (-1)^n \sin^{-1}(0)$, where n is an integer. Since $\sin^{-1}(0) = 0$, the general solution is $n\pi$.

When $n = 0$, we get 0, and this solution is in $[0, \pi]$.

When $n = 1$, we get π, and this solution is also in $[0, \pi]$.

When $n = 2$, we get 2π, and this solution is outside of $[0, \pi]$. It should be clear that for higher values of n, we will get solutions not in $[0, \pi]$.

When $n = -1$, we get $-\pi$, and this solution is outside of $[0, \pi]$. It should be clear that for lower values of n, we will get solutions not in $[0, \pi]$.

Therefore, there are two solutions of the equation $\sin \theta = 0$ in the interval $[0, \pi]$: 0 and π.

Basic Cosine Equation

The equation $\cos \theta = a$, where $-1 \le a \le 1$, is called the *basic cosine equation*. Again, we will use the unit circle to find the general solution of this equation.

Consider the unit circle, and let the points of intersection of the unit circle and the line $x = a$ be A and B, as shown in the following figure.

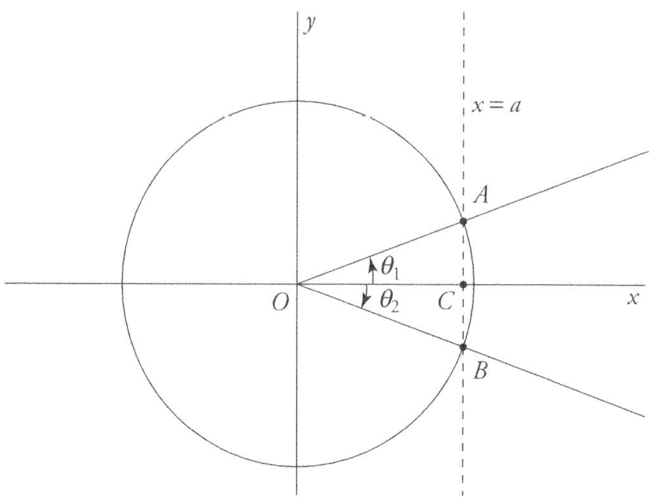

Let O be the origin and let C be the point of intersection of the line $x = a$ and the positive x-axis.

Let $\angle COA = \theta_1$ and $\angle COB = \theta_2$. Then, by *Theorem 2.6*, $\cos \theta_1 = a$ and $\cos \theta_2 = a$.

Therefore, θ_1 and θ_2 are solutions of the equation $\cos \theta = a$. The right triangles ΔAOC and ΔBOC are congruent by the *HL Theorem.*

Therefore, $\angle COA = \angle COB$. Since θ_2 is measured clockwise, $\theta_2 = -\theta_1$.

As a result, the following counterclockwise angles in standard position are solutions of the equation $\cos \theta = a$:

$$\theta_1, 2\pi - \theta_1, 2\pi + \theta_1, 4\pi - \theta_1, 4\pi + \theta_1, 6\pi - \theta_1, 6\pi + \theta_1, \ldots$$

Also, the following clockwise angles in standard position are solutions of the equation $\cos \theta = a$:

$$-\theta_1, -2\pi + \theta_1, -2\pi - \theta_1, -4\pi + \theta_1, -4\pi - \theta_1, -6\pi + \theta_1, -6\pi - \theta_1, \ldots$$

By inspection, the general solution of the equation $\cos \theta = a$ is $2n\pi \pm \theta_1$, where n is an integer.

$$\begin{aligned} \text{Since } \theta_1 \text{ is a solution of } \cos\theta &= a, \\ \text{we have } \cos\theta_1 &= a. \\ \text{Then } \cos^{-1}\left(\cos\theta_1\right) &= \cos^{-1} a. \\ \text{Since } 0 \le \theta_1 \le \pi,\ \cos^{-1}\left(\cos\theta_1\right) &= \theta_1. \\ \text{Therefore, } \theta_1 &= \cos^{-1} a. \end{aligned}$$

Therefore, the general solution of the equation $\cos \theta = a$ is $2n\pi \pm \cos^{-1} a$, where n is an integer.

Theorem 5.2: The general solution of $\cos \theta = a$ is $2n\pi \pm \cos^{-1} a$, where n is an integer and $-1 \le a \le 1$.

Example 4: Find the general solution of the equation $\cos\theta = \frac{1}{2}$, and find solutions of the equation in the interval $[0, \pi]$.

By *Theorem 5.2*, the general solution is $2n\pi \pm \cos^{-1}\left(\frac{1}{2}\right)$, where n is an integer. Since $\cos^{-1}\left(\frac{1}{2}\right) = \frac{\pi}{3}$, the general solution is $2n\pi \pm \frac{\pi}{3}$.

When $n = 0$, we get $\pm\frac{\pi}{3}$. The solution $\frac{\pi}{3}$ is in $[0, \pi]$.

When $n = 1$, we get $2\pi \pm \frac{\pi}{3}$, and both solutions are outside of $[0, \pi]$.

When $n = 2$, we get $4\pi \pm \frac{\pi}{3}$, and both solutions are outside of $[0, \pi]$. It should be clear that for higher values of n, we will get solutions not in $[0, \pi]$.

When $n = -1$, we get $-2\pi \pm \frac{\pi}{3}$, and neither solution is in $[0, \pi]$. It should be clear that for lower values of n, we will get solutions not in $[0, \pi]$.

Therefore, the only solution of the equation $\cos\theta = \frac{1}{2}$ in the interval $[0, \pi]$ is $\frac{\pi}{3}$.

Example 5: Find the general solution of the equation $\cos\theta = -\frac{\sqrt{3}}{2}$, and find solutions in the interval $[0, 2\pi]$.

By *Theorem 5.2*, the general solution is $2n\pi \pm \cos^{-1}\left(-\frac{\sqrt{3}}{2}\right)$, where n is an integer. Since $\cos^{-1}\left(-\frac{\sqrt{3}}{2}\right) = \frac{5\pi}{6}$, the general solution is $2n\pi \pm \frac{5\pi}{6}$.

When $n = 0$, we get $\pm\frac{5\pi}{6}$, and the solution $\frac{5\pi}{6}$ is in $[0, 2\pi]$.

When $n = 1$, we get $2\pi \pm \frac{5\pi}{6}$, and the solution $2\pi - \frac{5\pi}{6} = \frac{7\pi}{6}$ is in $[0, 2\pi]$. It should be clear that for higher values of n, we will get solutions not in $[0, 2\pi]$.

When $n = -1$, we get $-2\pi \pm \frac{5\pi}{6}$, and neither solution is in $[0, 2\pi]$. It should be clear that for lower values of n, we will get solutions not in

$[0, 2\pi]$. Therefore, the solutions of the equation $\cos\theta = -\frac{\sqrt{3}}{2}$ in the interval $[0, 2\pi]$ are $\frac{5\pi}{6}$ and $\frac{7\pi}{6}$.

Basic Tangent Equation

The equation $\tan\theta = a$, where a is a real number, is called the *basic tangent equation.* We will use the unit circle to find the general solution of this equation.

Without a loss of generality, assume $a > 0$. Suppose the solution of $\tan\theta = a$ in the first quadrant is θ_1. Let $P(x, y)$ be an arbitrary point on the terminal side of θ_1 other than the origin O. Then, by *Theorem 2.18*, $a = \frac{y}{x}$. That is, the point P is on the line $y = ax$. Suppose the line $y = ax$ intersects the unit circle in the first quadrant at $A(p, q)$.

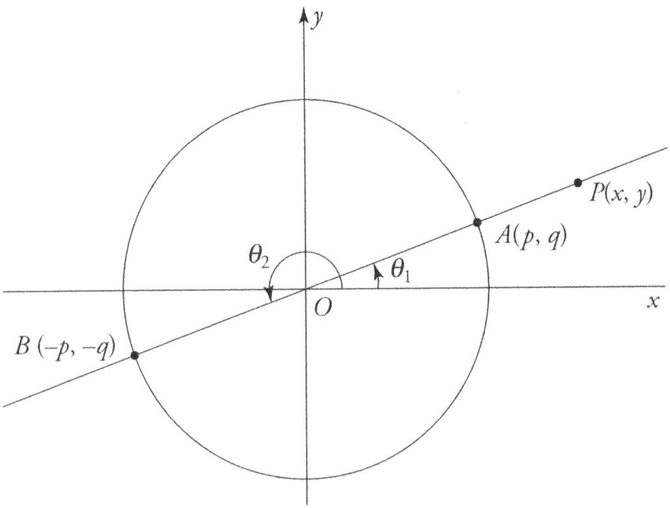

Then, since A is on the unit circle, $p^2 + q^2 = 1$, and $a = \frac{q}{p}$ since A is on the line $y = ax$. The point $B(-p, -q)$ is on the unit circle since $(-p)^2 + (-q)^2 = 1$, and B is on the line $y = ax$ since $\frac{-q}{-p} = \frac{q}{p} = a$. That is, B is the second point of intersection of the line $y = ax$ and the unit circle. Let θ_2 be the angle in standard position with terminal side equal to ray $\overrightarrow{OB}$.

Since $B(-p, -q)$ is on the terminal side of θ_2, by *Theorem 2.18*, $\tan\theta_2 = \frac{-q}{-p} = \frac{q}{p} = a$. That is, θ_2 is a solution of the equation $\tan\theta = a$.

Since $\angle BOA$ is a straight angle, $\theta_2 = \pi + \theta_1$.

Consequently, the following counterclockwise angles in standard position are solutions of the equation $\tan\theta = a$:

$$\theta_1, \pi + \theta_1, 2\pi + \theta_1, 3\pi + \theta_1, 4\pi + \theta_1, 5\pi + \theta_1, \ldots$$

Also, the following clockwise angles in standard position are solutions of the equation $\tan\theta = a$:

$$-\pi + \theta_1, -2\pi + \theta_1, -3\pi + \theta_1, -4\pi + \theta_1, -5\pi + \theta_1, \ldots$$

By inspection, the general solution of the equation $\tan\theta = a$ is $n\pi + \theta_1$, where n is an integer.

Since $-\frac{\pi}{2} < \theta_1 < \frac{\pi}{2}$, $\theta_1 = \tan^{-1}(a)$.

Theorem 5.3: The general solution of the equation $\tan\theta = a$ is $n\pi + \tan^{-1}(a)$, where n is an integer.

Example 6: Find the general solution of the equation $\tan\theta = 1$, and find solutions of the equation in the interval $[0, 2\pi]$.

By *Theorem 5.3*, the general solution is $n\pi + \tan^{-1}(1)$, where n is an integer. Since $\tan^{-1}(1) = \frac{\pi}{4}$, the general solution is $n\pi + \frac{\pi}{4}$.

When $n = 0$, we get $\frac{\pi}{4}$, and this solution is in $[0, 2\pi]$.

When $n = 1$, we get $\pi + \frac{\pi}{4} = \frac{5\pi}{4}$, and this solution is also in $[0, 2\pi]$.

When $n = 2$, we get $2\pi + \frac{\pi}{4}$, and this solution is outside of $[0, 2\pi]$. It should be clear that for higher values of n, we will get solutions not in $[0, 2\pi]$.

When $n = -1$, we get $-\pi + \frac{\pi}{4}$, and this solution is not in $[0, 2\pi]$. It should be clear that for lower values of n, we will get solutions not in $[0, 2\pi]$.

Therefore, there are two solutions of the equation $\tan\theta = 1$ in the interval $[0, 2\pi]$: $\frac{\pi}{4}$ and $\frac{5\pi}{4}$.

Example 7: Find the general solution of the equation $\tan\theta = -\sqrt{3}$, and find solutions in the interval $[0, \pi]$.

By *Theorem 5.3*, the general solution is $n\pi + \tan^{-1}\left(-\sqrt{3}\right)$, where n is an integer. Since $\tan^{-1}\left(-\sqrt{3}\right) = -\frac{\pi}{3}$, the general solution is $n\pi - \frac{\pi}{3}$.

When $n = 0$, we get $-\frac{\pi}{3}$, and this solution is not in $[0, \pi]$.

When $n = 1$, we get $\pi - \frac{\pi}{3} = \frac{2\pi}{3}$, and this solution is in $[0, \pi]$.

When $n = 2$, we get $2\pi - \frac{\pi}{3}$, and this solution is not in $[0, \pi]$. It should be clear that for higher values of n, we will get solutions not in $[0, \pi]$.

When $n = -1$, we get $-\pi - \frac{\pi}{3}$, and this solution is not in $[0, \pi]$. It should be clear that for lower values of n, we will get solutions not in $[0, \pi]$.

Therefore, the only solution of the equation $\tan\theta = -\sqrt{3}$ in the interval $[0, \pi]$ is $\frac{2\pi}{3}$.

Example 8: Find the general solution of the equation $\tan\theta = 0$, and find solutions in the interval $[0, \pi]$.

By *Theorem 5.3*, the general solution is $n\pi + \tan^{-1}(0)$, where n is an integer. Since $\tan^{-1}(0) = 0$, the general solution is $n\pi$.

When $n = 0$, we get 0, and this solution is in $[0, \pi]$.

When $n = 1$, we get π, and this solution is also in $[0, \pi]$.

When $n = 2$, we get 2π, and this solution is outside of $[0, \pi]$. It should be clear that for higher values of n, we will get solutions not in $[0, \pi]$.

When $n = -1$, we get $-\pi$, and this solution is outside of $[0, \pi]$. It should be clear that for lower values of n, we will get solutions not in $[0, \pi]$.

Therefore, there are two solutions of the equation $\tan \theta = 0$ in the interval $[0, \pi]$: 0 and π.

Chapter Check-Out

Questions

1. Find the solutions of $\sin\theta = -\frac{1}{2}$ in the interval $[0, 2\pi]$.
2. Find the solutions of $\cos\theta = -\frac{1}{2}$ in the interval $[0, 2\pi]$.
3. Find the solutions of $\tan \theta = -1$ in the interval $[0, 2\pi]$.
4. True or False: $\cos^{-1}\left(\frac{1}{3}\right)$ is a solution of the equation $\cos\theta = \frac{1}{3}$.
5. True or False: $\sin^{-1}\left(\frac{1}{2}\right)$ is a solution of the equation $\sin\theta = \frac{1}{2}$.
6. True or False: $\pi + \tan^{-1}(2)$ is a solution of the equation $\tan \theta = 2$.

Answers

1. $\frac{7\pi}{6}, \frac{11\pi}{6}$
2. $\frac{2\pi}{3}, \frac{4\pi}{3}$
3. $\frac{3\pi}{4}, \frac{7\pi}{4}$
4. True
5. True
6. True

Chapter 6

TRIGONOMETRY OF TRIANGLES

Chapter Check-In

- ❑ Solving right triangles
- ❑ Solving triangles using the Law of Sines
- ❑ Solving triangles using the Law of Cosines
- ❑ Finding the area of triangles

Common Core Standard: Apply trigonometry to general triangles

Derive the formula $A = \frac{1}{2}ab\sin(C)$ for the area of a triangle by drawing an auxiliary line from a vertex perpendicular to the opposite side. (HSG.SRT.D.9)

Prove the Laws of Sines and Cosines and use them to solve problems. (HSG.SRT.D.10)

Understand and apply the Law of Sines and the Law of Cosines to find unknown measurements in right and non-right triangles. (HSG.SRT.D.11)

Use trigonometric ratios and the Pythagorean Theorem to solve right triangles in applied problems. (HSG.SRT.C.8)

Please also refer to the standards for mathematical practice, found here: www.corestandards.org/Math/Practice/.

We will adopt the following convention for a given triangle ΔABC for convenience. The measure of $\angle A$ is α, the measure of $\angle B$ is β, and the measure of $\angle C$ is γ. The length of the side facing $\angle A$ is a, the length of the side facing $\angle B$ is b, and the length of the side facing

$\angle C$ is c. Finding the six numbers α, β, γ, a, b, and c of a given triangle is known as *solving the triangle.*

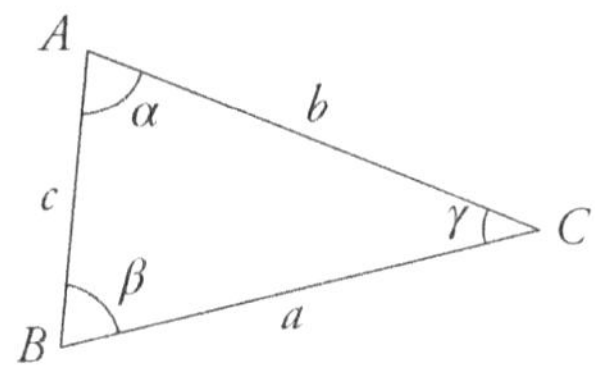

We may also need some theorems from geometry to solve triangles. (See *CliffsNotes Geometry Common Core Quick Review* for proofs of the following theorems.)

Trigonometry Theorems

The trigonometry theorems in this book are numbered for organizational purposes. For example, *Theorem 2.2* is the second theorem given in Chapter 2. Study these theorems by their content, not by their number, as the theorem numbering has no significance outside of this book.

*Theorem 6.1 (**Pythagorean Theorem**):* If ΔABC is a right triangle so that $\angle C$ is the right angle, then $a^2 + b^2 = c^2$.

*Theorem 6.2 (**Converse of the Pythagorean Theorem**):* If ΔABC is a triangle so that $a^2 + b^2 = c^2$, then ΔABC is a right triangle and $\angle C$ is the right angle.

Theorem 6.3: In a triangle, a side facing a larger angle is longer than a side facing a smaller angle.

Theorem 6.4: In a triangle, an angle facing a longer side is larger than an angle facing a shorter side.

Theorem 6.5 (Triangle Inequality Theorem): In a triangle, the sum of any two sides is longer than the third side.

Theorem 6.6 (SAS Theorem for Congruence): Given two triangles ΔABC and ΔDEF so that $\angle A = \angle D$, $AB = DE$, and $AC = DF$, then the triangles are congruent.

You want to look at the statement of the SAS Theorem in the following way. If the lengths of two sides and the measure of the included angle of those two sides are known of a triangle, then there is only one such triangle in the sense, that if there is another triangle with the same properties then the two triangles are congruent. That is, if there are two triangles with the same properties, then we can "move" one of those triangles on top of the other triangle. In other words, if there are two triangles with the given properties of the SAS Theorem, then they are copies of the same triangle.

Theorem 6.7 (SAA Theorem for Congruence): Given two triangles ΔABC and ΔDEF so that $AC = DF$, $\angle A = \angle D$, and $\angle B = \angle E$, then the triangles are congruent.

That is, if the length of a side and the measures of two angles are known of a triangle, then there is only one such triangle in the sense, that if there is another triangle with the same properties then the two triangles are congruent.

Theorem 6.8 (SSS Theorem for Congruence): Two triangles with three equal sides are congruent.

That is, if the lengths of all three sides are known of a triangle, then there is only one such triangle in the sense, that if there is another triangle with the same properties then the two triangles are congruent.

Theorem 6.9 (HL Theorem for Congruence): If two right triangles have equal hypotenuses and one pair of equal legs, then the two triangles are congruent.

That means, if the length of the hypotenuse and the length of one side are known of a right triangle, then there is only one such triangle in the sense, that if there is another triangle with the same properties then the two triangles are congruent.

Theorem 6.10: The sum of the degrees of angles of a triangle is 180°.

First, we will find ways to solve right triangles. Then we can use that knowledge to find methods to solve any triangle.

Solving Right Triangles

Theorem 2.3 is very useful in solving right triangles. We can restate *Theorem 2.3* as follows.

Theorem 6.11: Let θ be an acute angle of a right triangle and suppose the length of the hypotenuse is r. Then the length of the adjacent side of θ is $r\cos\theta$ and the length of the opposite side of θ is $r\sin\theta$.

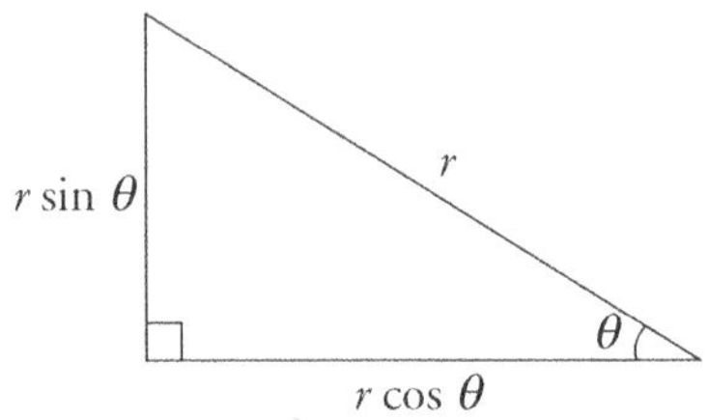

Let ΔABC be a right triangle so that $\angle C$ is the right angle.

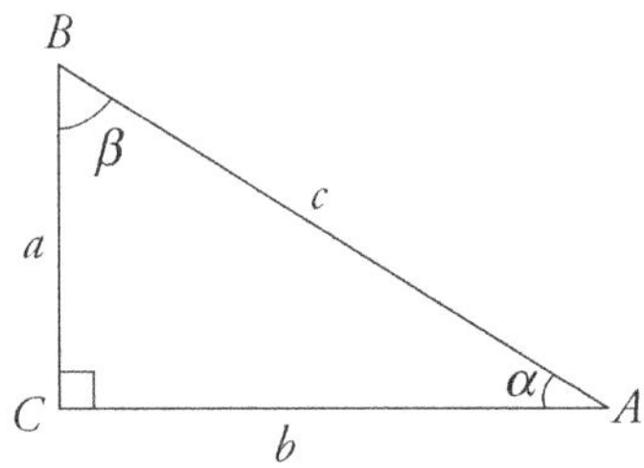

Then by *Theorem 6.11*, $a = c\sin\alpha$ and $b = c\cos\alpha$. Then $\frac{a}{b} = \frac{c\sin\alpha}{c\cos\alpha} = \tan\alpha$. Also, $b = c\sin\beta$ and $a = c\cos\beta$. That is, $\cos\alpha = \sin\beta$ and $\sin\alpha = \cos\beta$. Then $\tan\beta = \frac{b}{a}$.

Example 1: Let ΔABC be a right triangle so that $\angle C$ is the right angle, $\beta = 20°$, and $b = 16$. Solve the triangle if possible.

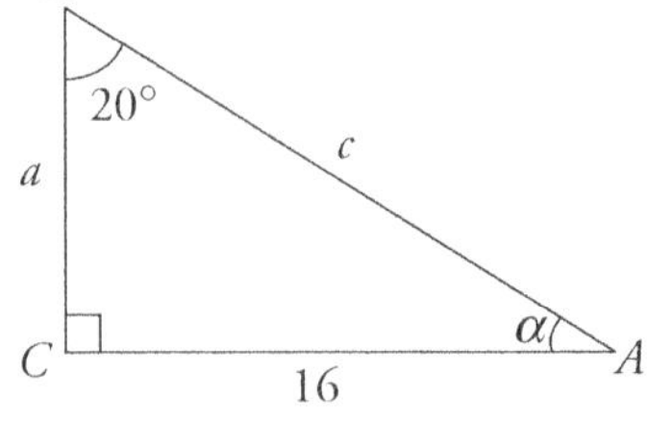

By the SAA Theorem, we know that there is one such triangle. By *Theorem 6.9*,

$$90° + 20° + \alpha = 180°$$
$$\alpha = 70°$$

By *Theorem 6.11*, $c \sin 20° = 16$. Therefore, $c = \dfrac{16}{\sin 20°}$.

By *Theorem 6.11*, $\tan 20° = \dfrac{16}{a}$. Therefore, $a = \dfrac{16}{\tan 20°}$.

Therefore, we have solved the given triangle. The following are the six numbers:

$$\alpha = 70°,\ \beta = 20°,\ \gamma = 90°,\ a = \frac{16}{\tan 20°},\ b = 16,\ c = \frac{16}{\sin 20°}$$

We can find approximate values of a and c using a scientific calculator. The following are the six numbers rounded off to two decimal places:

$$\alpha = 70°,\ \beta = 20°,\ \gamma = 90°,\ a \approx 43.96,\ b = 16,\ c \approx 46.78$$

Example 2: Let ΔABC be a right triangle so that $\angle C$ is the right angle, $a = 13$, and $b = 8$. Solve the triangle if possible.

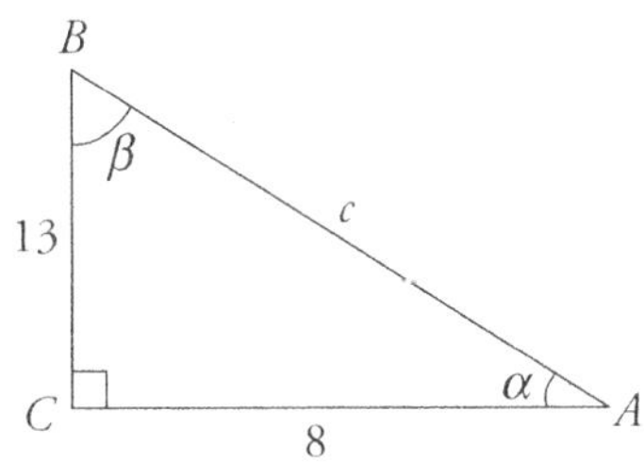

By the SAS Theorem, there is one such triangle.

By the Pythagorean Theorem,

$$c^2 = 8^2 + 13^2$$
$$c = \sqrt{233}$$

By *Theorem 6.11*,

$$c\cos\beta = 13$$

$$\cos\beta = \frac{13}{\sqrt{233}}$$

Since β is acute, β is in the interval (0°, 90°), and there is only one solution to the above basic cosine equation. That is, $\beta = \cos^{-1}\left(\frac{13}{\sqrt{233}}\right)$.

By *Theorem 6.11* again,

$$c\cos\alpha = 8$$

$$\cos\alpha = \frac{8}{\sqrt{233}}$$

Since α is acute, $\alpha = \cos^{-1}\left(\frac{8}{\sqrt{233}}\right)$.

Therefore, $\alpha = \cos^{-1}\left(\frac{8}{\sqrt{233}}\right)$, $\beta = \cos^{-1}\left(\frac{13}{\sqrt{233}}\right)$, $\gamma = 90°$, $a = 13$, $b = 8$, and $c = \sqrt{233}$.

If we round off all angles to the nearest degree and all lengths to the nearest hundredth, then $\alpha = 58°$, $\beta = 32°$, $\gamma = 90°$, $a = 13$, $b = 8$, and $c \approx 15.26$.

In many applications, certain angles are referred to by special names. Two of these special names are **angle of elevation** and **angle of depression.** The examples shown in the following figure make use of these terms.

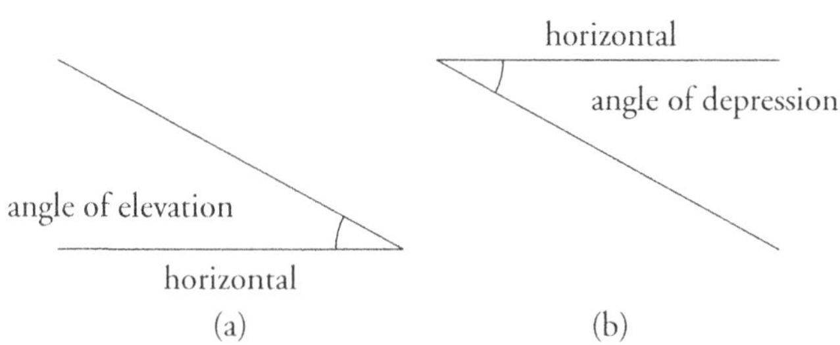

Example 3: A large airplane (plane A) flying at 26,000 feet sights a smaller plane (plane B) traveling at an altitude of 24,000 feet. The angle of depression is 40°. What is the line of sight distance between the two planes? Round off your answer to the nearest foot.

Let the line of sight distance between the planes be c feet. The vertical height between the positions of the planes is 26,000 – 24,000 = 2,000 feet, as shown in the following figure.

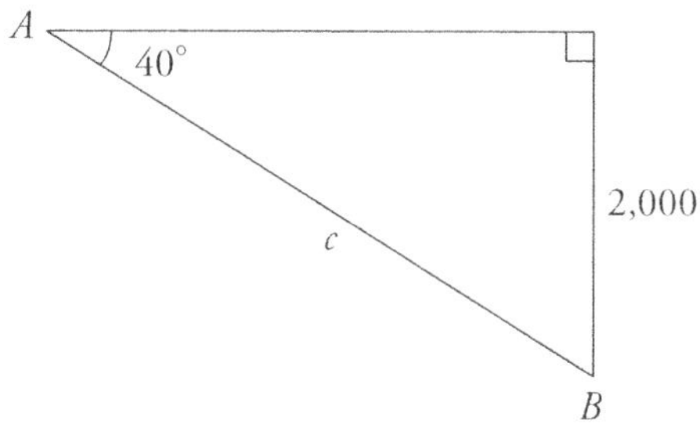

By *Theorem 6.11*, $c \sin 40° = 2{,}000$. Therefore, $c = \dfrac{2{,}000}{\sin 40°}$ (that's approximately 3,111 feet rounded to the nearest foot).

Example 4: A ladder must reach the top of a building. The base of the ladder will be 25 feet from the base of the building. The angle of elevation from the base of the ladder to the top of the building is 64°. Find the height of the building and the length of the ladder. Round off your answers to the nearest foot.

Let the height of the building be h feet and the length of the ladder be m feet. The following figure depicts the problem.

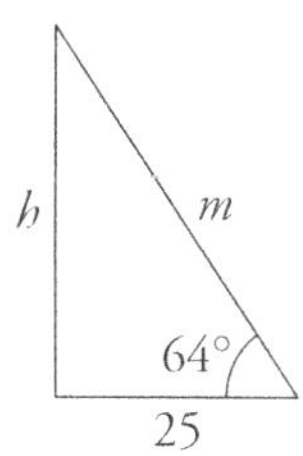

By *Theorem 6.11*, $m \cos 64° = 25$. Therefore, $m = \dfrac{25}{\cos 64°} \approx 57$.

By *Theorem 6.11*, $\tan 64° = \dfrac{h}{25}$. Therefore, $h = 25 \tan 64° \approx 51$.

The height of the building is approximately 51 feet, and the length of the required ladder is a little over 57 feet. To be safe, the length of the ladder should be at least 58 feet.

Example 5: A woodcutter wants to determine the height of a tall tree. He stands at some distance from the tree and determines that the angle of elevation to the top of the tree is 40°. He moves 30 feet closer to the tree, and now the angle of elevation is 50°. If the woodcutter's eyes are 5 feet above the ground, how tall is the tree? Round off your answer to the nearest foot.

Let h be the height of the tree in feet from the eye level of the woodcutter, and let the distance from the point where the woodcutter was standing when the elevation angle is 50° to the base of the tree be x feet. The following figure can help you visualize the problem.

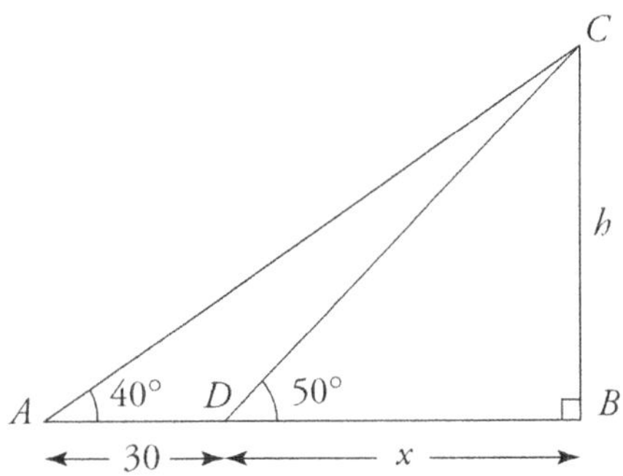

From the right triangle ΔCBD and by *Theorem 6.11*,

$$\tan 50° = \frac{h}{x}$$

$$h = x \tan 50° \qquad (1)$$

From the right triangle ΔCBA and by *Theorem 6.11*,

$$\tan 40° = \frac{h}{x+30}$$

$$h = (x+30)\tan 40° \qquad (2)$$

From equations (1) and (2), $x \tan 50° = (x + 30) \tan 40°$.

Solve the equation for x:

$$\begin{aligned} x\tan 50^\circ &= (x+30)\tan 40^\circ \\ x\tan 50^\circ &= x\tan 40^\circ + 30\tan 40^\circ \\ x(\tan 50^\circ - \tan 40^\circ) &= 30\tan 40^\circ \\ x &= \frac{30\tan 40^\circ}{(\tan 50^\circ - \tan 40^\circ)} \end{aligned}$$

Then by equation (1), $h = \dfrac{30\tan 40^\circ \tan 50^\circ}{(\tan 50^\circ - \tan 40^\circ)} \approx 85.069$.

Note that 5 feet must be added to the value of h to get the height of the tree because the woodcutter's eyes are 5 feet above the ground. Therefore, the tree is approximately 85 + 5 = 90 feet tall rounded to the nearest foot.

Solving Triangles: SSS and SAS Cases

The previous section covered solving right triangles. In this section and the next, we will find methods to solve *any* triangle. As we will discover soon, if we know some of these six numbers—α, β, γ, a, b, c—then we can find the other numbers. First, we consider the cases where the lengths of all three sides are known (SSS case) and where the lengths of two sides and the included angle are known (SAS case). With the addition of the following theorem, we can solve SSS and SAS cases.

*Theorem 6.12 (**Law of Cosines**):* Let ΔABC be a triangle with the adopted convention. Then,

$$\begin{aligned} a^2 &= b^2 + c^2 - 2bc\cos\alpha \\ b^2 &= a^2 + c^2 - 2ac\cos\beta \\ c^2 &= a^2 + b^2 - 2ab\cos\gamma \end{aligned}$$

Proof: We will prove the statement $c^2 = a^2 + b^2 - 2ab\cos\gamma$. The proofs of the other two statements are similar. Introduce a coordinate system and move ΔABC so that point C is at the origin and point B is on the positive x-axis. Then the coordinates of point B are $(a, 0)$. Let the coordinates of point A be (x, y), as shown in the following figure. The following proof does not depend on the position of point A. That is, the proof is valid if point A is in any of the quadrants, or if point A is on the y-axis.

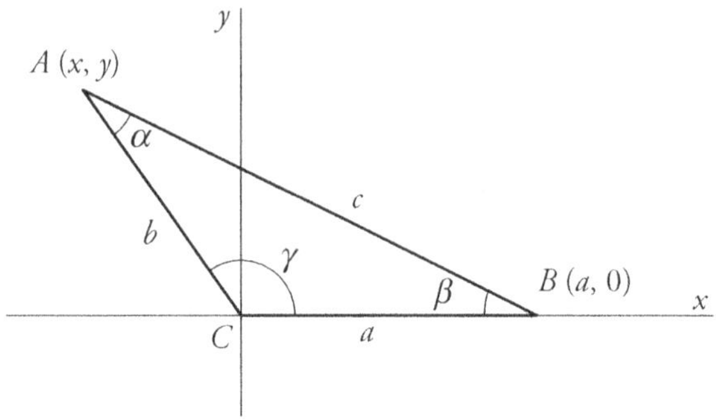

By the definitions of sin γ and cos γ (see Chapter 2), $\sin\gamma = \frac{y}{b}$ and $\cos\gamma = \frac{x}{b}$. Thus, the coordinates of A are $x = b\cos\gamma$ and $y = b\sin\gamma$. Using the distance formula,

$$c^2 = (y-0)^2 + (x-a)^2$$
$$c^2 = (b\sin\gamma - 0)^2 + (b\cos\gamma - a)^2$$
$$c^2 = b^2\sin^2\gamma + b^2\cos^2\gamma - 2ab\cos\gamma + a^2$$
$$c^2 = b^2(\sin^2\gamma + \cos^2\gamma) - 2ab\cos\gamma + a^2$$
$$c^2 = a^2 + b^2 - 2ab\cos\gamma$$

In the preceding formula, if γ is 90°, then cos 90° = 0, yielding the Pythagorean Theorem for right triangles.

Example 6: Let $a = 3$, $b = 4$, and $c = 7$ in ΔABC. Solve this triangle, if possible.

Since $a + b = c$, these numbers violate the triangle inequality. Therefore, there is no such triangle.

Example 7: Let $a = 3$, $b = 4$, and $c = 5$ in ΔABC. Solve this triangle, if possible.

Since $a^2 + b^2 = c^2$, by the Converse of the Pythagorean Theorem, ΔABC is a right triangle with $\gamma = 90°$. Then by the SSS Theorem for Congruence, there is only one such triangle.

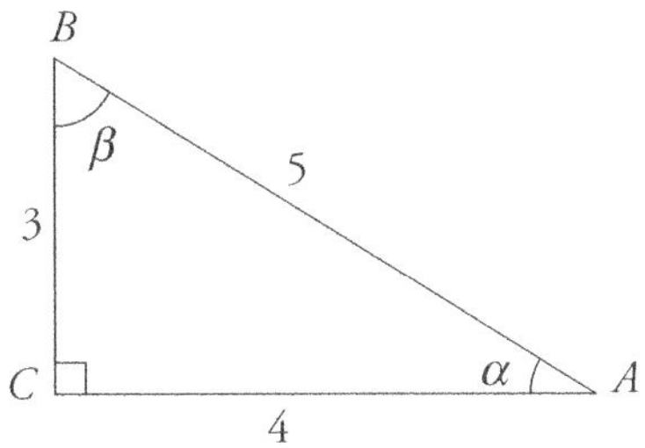

By *Theorem 6.11*, $\tan\alpha = \frac{3}{4}$. Since α is an acute angle, this equation has only one solution: $\alpha = \tan^{-1}\left(\frac{3}{4}\right)$

By *Theorem 6.10*, $\beta = 90° - \tan^{-1}\left(\frac{3}{4}\right)$. Therefore, $\alpha = \tan^{-1}\left(\frac{3}{4}\right)$, $\beta = 90° - \tan^{-1}\left(\frac{3}{4}\right)$, $\gamma = 90°$, $a = 3$, $b = 4$, and $c = 5$.

If we round off all angles to the nearest degree, then $\alpha \approx 37°$, $\beta \approx 53°$, $\gamma = 90°$, $a = 3$, $b = 4$, and $c = 5$.

Example 8: Let $a = 12$, $b = 7$, and $c = 6$ in ΔABC. Solve this triangle, if possible.

The triangle inequality holds, and since $12^2 \neq 7^2 + 6^2$, ΔABC is not a right triangle.

By the Law of Cosines Theorem,

$$b^2 = a^2 + c^2 - 2ac\cos\beta$$

Solve the above equation for $\cos\beta$.

$$\cos\beta = \frac{a^2 + c^2 - b^2}{2ac}$$

$$\cos\beta = \frac{12^2 + 6^2 - 7^2}{2(12)(6)}$$

$$\cos\beta = \frac{131}{144}$$

Since β is an angle of a triangle, β is in the interval (0°, 180°). In this interval, the above equation has only one solution: $\beta = \cos^{-1}\left(\frac{131}{144}\right)$, meaning $\beta \approx 24.53°$.

The measure of α can be found in a similar way.

$$a^2 = b^2 + c^2 - 2bc\cos\alpha$$

Solve the above equation for cos α.

$$\cos\alpha = \frac{b^2 + c^2 - a^2}{2bc}$$

$$\cos\alpha = \frac{7^2 + 6^2 - 12^2}{2(7)(6)}$$

$$\cos\alpha = -\frac{59}{84}$$

Since α is an angle of a triangle, α is in the interval (0°, 180°). In this interval, the above equation has only one solution: $\alpha = \cos^{-1}\left(-\frac{59}{84}\right)$, meaning $\alpha \approx 134.62°$.

Now by *Theorem 6.10*, $\gamma = 180° - \cos^{-1}\left(-\frac{59}{84}\right) - \cos^{-1}\left(\frac{131}{144}\right)$. If we round off the angles to the nearest degree, then the solution to this problem is:

$$\alpha \approx 135°,\ \beta \approx 25°,\ \gamma \approx 20°,\ a = 12,\ b = 7,\ c = 6$$

Next, consider a triangle with the lengths of two sides and the measure of the included angle known. Then by the SAS Theorem, there is only one such triangle.

Example 9: Let ΔABC be a triangle so that $a = 11$, $c = 10$, and $\beta = 71°$. Solve the triangle, if possible.

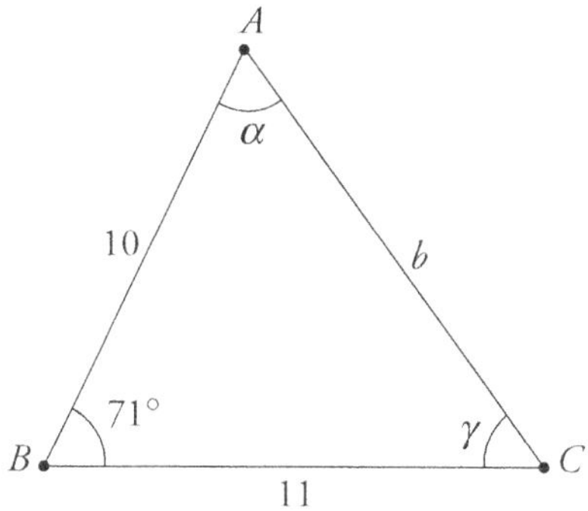

By the Law of Cosines Theorem,

$$b^2 = a^2 + c^2 - 2ac\ \cos\beta$$
$$b^2 = 11^2 + 10^2 - 2(11)(10)\cos 71^\circ$$
$$b^2 = 221 - 220\cos 71^\circ$$
$$b = \sqrt{221 - 220\cos 71^\circ}$$

By the Law of Cosines Theorem,

$$a^2 = b^2 + c^2 - 2bc\cos\alpha$$
$$11^2 = b^2 + 10^2 - 2(b)(10)\cos\alpha$$
$$\cos\alpha = \frac{b^2 + 10^2 - 11^2}{20b}$$
$$\cos\alpha = \frac{b^2 - 21}{20b}$$

Since α is an angle of a triangle, α is in the interval $(0^\circ, 180^\circ)$. In this interval, the above equation has only one solution: $\alpha = \cos^{-1}\left(\frac{b^2 - 21}{20b}\right)$, meaning $\alpha \approx 58^\circ$.

Then the remaining angle, γ, can be found using *Theorem 6.10.*

$$\gamma = 180^\circ - 71^\circ - \cos^{-1}\left(\frac{b^2 - 21}{20b}\right)$$

If we round off all angles to the nearest degree and all lengths to the nearest hundredth, then the solution is:

$$\alpha \approx 58°, \beta = 71°, \gamma \approx 51°, a = 11, b \approx 12.22, c = 10$$

Solving Triangles: SAA and SSA Cases

The Law of Cosines discussed in the last section to solve general triangles *cannot* be used in the SAA and SSA cases. Consider the following problem.

Example 10: Let ABC be a triangle so that $\beta = 71°$, $\gamma = 51°$, and $b = 12$. Solve the triangle, if possible.

By the SAA Congruence Theorem, we know there is only one such triangle. We can find α by using *Theorem 6.10*.

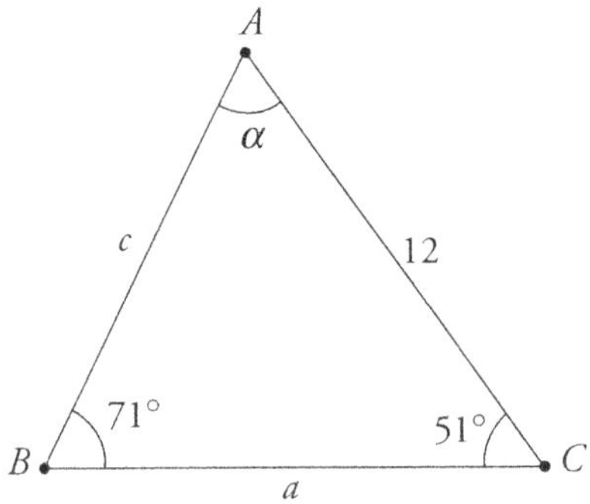

However, we cannot use $b^2 = a^2 + c^2 - 2ac\cos\beta$ because we do not know both a and c. We cannot use $a^2 = b^2 + c^2 - 2bc\cos\alpha$ because we do not know both a and c, and we cannot use $c^2 = a^2 + b^2 - 2ab\cos\gamma$ for the same reason.

Therefore, we cannot solve this triangle by using known theorems. We need another theorem.

*Theorem 6.13 (**Law of Sines**):* Let ΔABC be a triangle with the adopted convention. Then, $\dfrac{a}{\sin\alpha} = \dfrac{b}{\sin\beta} = \dfrac{c}{\sin\gamma}$.

There are three parts to this theorem in the given condensed version:

$$\frac{a}{\sin\alpha} = \frac{b}{\sin\beta}, \quad \frac{b}{\sin\beta} = \frac{c}{\sin\gamma}, \text{ and } \frac{a}{\sin\alpha} = \frac{c}{\sin\gamma}$$

Proof: There are two cases; the triangle can be either an acute triangle or the triangle can be an obtuse triangle. We will prove the first case; the proof of the other case is similar.

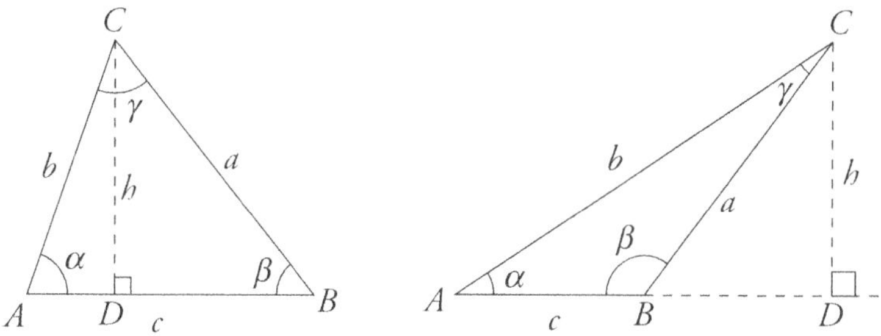

(a) (b)

Case 1: Assume the triangle is acute. Draw a perpendicular from point C to line $\overleftrightarrow{AB}$, and let point D be the foot of the perpendicular. Then point D lies between points A and B, as shown in figure (a) above. Let the length of segment $\overline{CD}$ be h. By using *Theorem 6.11* on right triangle ΔACD, we get:

$$h = b \sin \alpha \quad (1)$$

By applying *Theorem 6.11* on right triangle ΔBCD, we get:

$$h = a \sin \beta \quad (2)$$

By (1) and (2),

$$b \sin \alpha = a \sin \beta$$

Divide both sides of the above equation by $\sin \alpha \sin \beta$ and you get:

$$\frac{b}{\sin \beta} = \frac{a}{\sin \alpha}$$

Case 2: Assume the triangle is obtuse and assume that $\angle B$ is the obtuse angle. Draw a perpendicular from point C to line $\overleftrightarrow{AB}$, and let point D be the foot of the perpendicular. Then point D lies on the extended side $\overline{AB}$, as shown in figure (b) above. Let the length of segment $\overline{CD}$ be h. By using *Theorem 6.11* on right triangle ΔACD, we get:

$$h = b \sin \alpha \quad (3)$$

By applying *Theorem 6.11* on right triangle ΔBCD, we get:

$$h = a \sin (180^\circ - \beta) \quad (4)$$

By *Theorem 2.11*, $\sin (180^\circ - \beta) = \sin \beta$. Therefore, (4) can be written as:

$$h = a \sin \beta \quad (5)$$

By (3) and (5),

$$b \sin \alpha = a \sin \beta$$

Divide both sides of the above equation by $\sin \alpha \sin \beta$ and you get:

$$\frac{b}{\sin \beta} = \frac{a}{\sin \alpha}$$

Let us see if we can solve the triangle in Example 10 by using the Law of Sines Theorem. Example 11 that follows is a restatement of Example 10.

Example 11: Let ΔABC be a triangle so that $\beta = 71^\circ$, $\gamma = 51^\circ$, and $b = 12$. Solve the triangle, if possible.

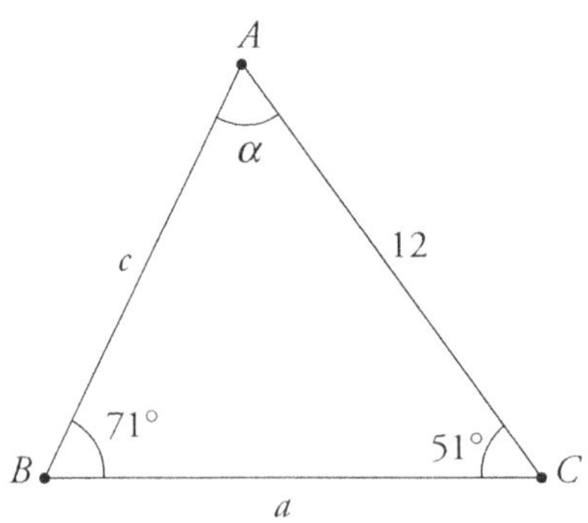

By *Theorem 6.10*, $\alpha = 58°$.

By the Law of Sines Theorem,

$$\frac{c}{\sin\gamma} = \frac{b}{\sin\beta}$$

$$\frac{c}{\sin 51°} = \frac{12}{\sin 71°}$$

$$c = \frac{12\sin 51°}{\sin 71°}$$

By applying the Law of Sines Theorem again,

$$\frac{a}{\sin\alpha} = \frac{b}{\sin\beta}$$

$$\frac{a}{\sin 58°} = \frac{12}{\sin 71°}$$

$$a = \frac{12\sin 58°}{\sin 71°}$$

If we round off the lengths of the segments to the nearest hundredth, the solution is:

$$\alpha = 58°,\ \beta = 71°,\ \gamma = 51°,\ a \approx 10.76,\ b = 12,\ c \approx 9.86$$

Example 12: Let ΔABC be a triangle so that $\beta = 121°$, $c = 13$, and $b = 12$. Solve the triangle, if possible.

The triangle, if it exists, is an obtuse triangle with $\angle B$ being the largest angle. But $c > b$, which violates the statement of *Theorem 6.3*. Therefore, there is no such triangle.

Example 13: Let ΔABC be a triangle so that $\alpha = 125°$, $\beta = 35°$, and $b = 42$. Solve the triangle, if possible.

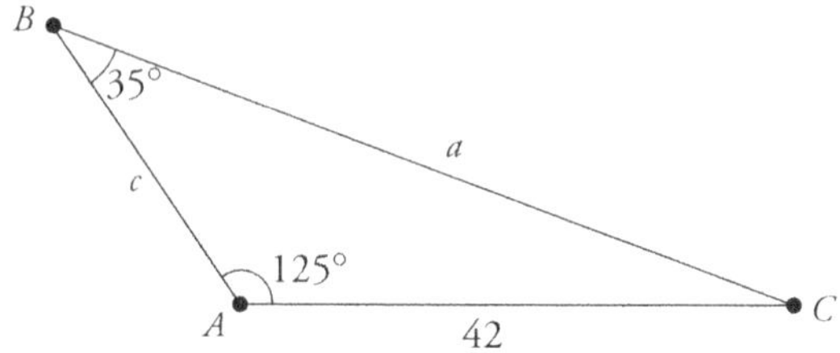

By the SAA Congruence Theorem, we know there is only one such triangle.

By *Theorem 6.10*, $\gamma = 20°$.

By the Law of Sines Theorem,

$$\frac{a}{\sin\alpha} = \frac{b}{\sin\beta}$$

$$\frac{a}{\sin 125°} = \frac{42}{\sin 35°}$$

$$a = \frac{42\sin 125°}{\sin 35°}$$

By the Law of Sines Theorem again,

$$\frac{c}{\sin\gamma} = \frac{b}{\sin\beta}$$

$$\frac{c}{\sin 20°} = \frac{42}{\sin 35°}$$

$$c = \frac{42\sin 20°}{\sin 35°}$$

If we round off the lengths of the segments to the nearest hundredth, the solution is:

$$\alpha = 125°,\ \beta = 35°,\ \gamma = 20°,\ a \approx 59.98,\ b = 42,\ c \approx 25.04$$

Now consider the case where we know the lengths of two sides of a triangle and the measure of the angle opposite one of them. This case is known as SSA. Unfortunately, there is no congruence theorem for this case. In other words, if a triangle exists with the given properties, then there is no guarantee that it is the only such triangle. Therefore, we have to be extremely judicious when handling problems of this kind.

Example 14: Let ΔABC be a triangle so that $a = 13$, $b = 20$, and $\alpha = 35°$. Solve the triangle, if possible.

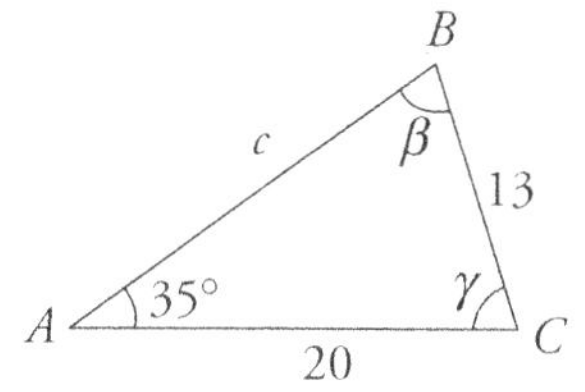

The following figure incorporates the given information, if there is such a triangle.

By the Law of Sines Theorem,

$$\frac{a}{\sin\alpha} = \frac{20}{\sin\beta}$$

$$\frac{13}{\sin 35°} = \frac{20}{\sin\beta}$$

$$\sin\beta = \frac{20\sin 35°}{13}$$

This is a basic sine equation if $\frac{20\sin 35°}{13} < 1$. You can use a calculator to check this: $\frac{20\sin 35°}{13}$ is approximately 0.88. Since β is the measure of an angle of a triangle, β is in the interval (0°, 180°). There are two solutions to the basic sine equation in this interval: $\beta = \sin^{-1}\left(\frac{20\sin 35°}{13}\right)$ and $\beta = 180° - \sin^{-1}\left(\frac{20\sin 35°}{13}\right)$.

Case 1: Suppose $\beta = \sin^{-1}\left(\frac{20\sin 35°}{13}\right)$.

If we round off β to the nearest degree, then $\beta \approx 62°$. Then by *Theorem 6.10*, $\gamma \approx 83°$.

By using the Law of Sines Theorem again,

$$\frac{c}{\sin\gamma} = \frac{a}{\sin\alpha}$$

$$\frac{c}{\sin 83^\circ} \approx \frac{13}{\sin 35^\circ}$$

$$c \approx \frac{13\sin 83^\circ}{\sin 35^\circ} \approx 22.5$$

Therefore, we have one solution to this problem.

$$\alpha = 35^\circ,\ \beta \approx 62^\circ,\ \gamma \approx 83^\circ,\ a = 13,\ b = 20,\ c \approx 22.5$$

Case 2: Suppose $\beta = 180^\circ - \sin^{-1}\left(\frac{20\sin 35^\circ}{13}\right)$.

If we round off β to the nearest degree, then $\beta \approx 118^\circ$. Then by *Theorem 6.10*, $\gamma \approx 27^\circ$.

By using the Law of Sines Theorem again,

$$\frac{c}{\sin\gamma} = \frac{a}{\sin\alpha}$$

$$\frac{c}{\sin 27^\circ} \approx \frac{13}{\sin 35^\circ}$$

$$c \approx \frac{13\sin 27^\circ}{\sin 35^\circ} \approx 10.29$$

Therefore, we have another solution to this problem.

$$\alpha = 35^\circ,\ \beta \approx 118^\circ,\ \gamma \approx 27^\circ,\ a = 13,\ b = 20,\ c \approx 10.29$$

This problem has two solutions. That means there are two triangles with the given properties. In the following figure, ΔAB_1C is the first solution and ΔAB_2C is the second solution.

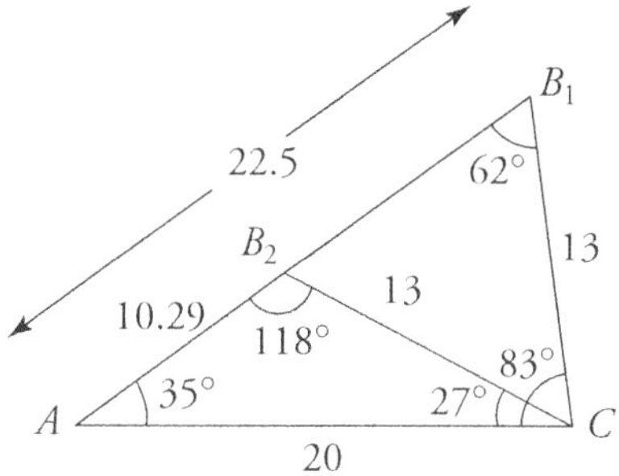

Example 15: Let ΔABC be a triangle so that $a = 22$, $b = 20$, and $\alpha = 35°$. Solve the triangle, if possible.

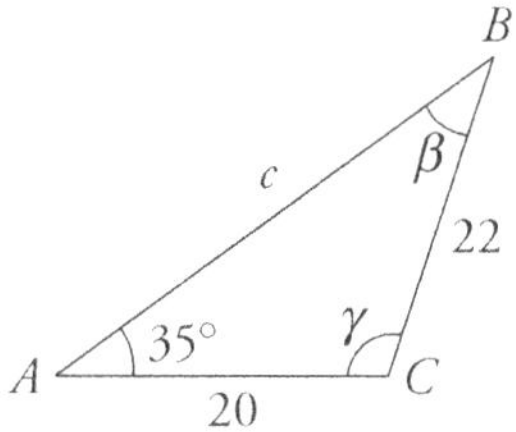

By the Law of Sines Theorem,

$$\frac{a}{\sin\alpha} = \frac{b}{\sin\beta}$$

$$\frac{22}{\sin 35°} = \frac{20}{\sin\beta}$$

$$\sin\beta = \frac{10\sin 35°}{11}$$

Since $\frac{10\sin 35°}{11} < 1$, the equation above is a basic sine equation. Since β is an angle measure of an angle of a triangle, it is in the interval (0°, 180°). Therefore, the above equation has two solutions: $\sin^{-1}\left(\frac{10\sin 35°}{11}\right)$ and $180° - \sin^{-1}\left(\frac{10\sin 35°}{11}\right)$.

Case 1: Suppose $\beta = \sin^{-1}\left(\frac{10\sin 35^\circ}{11}\right)$.

To the nearest degree, $\beta \approx 31^\circ$. Then by *Theorem 6.11*, $\gamma \approx 114^\circ$.

Again, by the Law of Sines Theorem,

$$\frac{c}{\sin\gamma} = \frac{a}{\sin\alpha}$$

$$\frac{c}{\sin 114^\circ} = \frac{22}{\sin 35^\circ}$$

$$c = \frac{22\sin 114^\circ}{\sin 35^\circ} \approx 35.04$$

Therefore, one solution to this problem is:

$$\alpha = 35^\circ,\ \beta \approx 31^\circ,\ \gamma \approx 114^\circ,\ a = 22,\ b = 20,\ c \approx 35.04$$

Case 2: Suppose $\beta = 180^\circ - \sin^{-1}\left(\frac{10\sin 35^\circ}{11}\right)$.

In this case, $\beta \approx 149^\circ$.

You should notice a problem right away. If $\beta \approx 149^\circ$, then it violates *Theorem 6.4* because $a > b$, but 35° is not larger than 149°. However, if you do not notice this, then you will soon realize that $35^\circ + 149^\circ > 180^\circ$; this result violates *Theorem 6.10*. Therefore, Case 2 does not lead to a solution, and this problem has only one solution.

Example 16: Let ΔABC be a triangle so that $a = 30$, $b = 20$, and $\alpha = 135^\circ$. Solve the triangle, if possible.

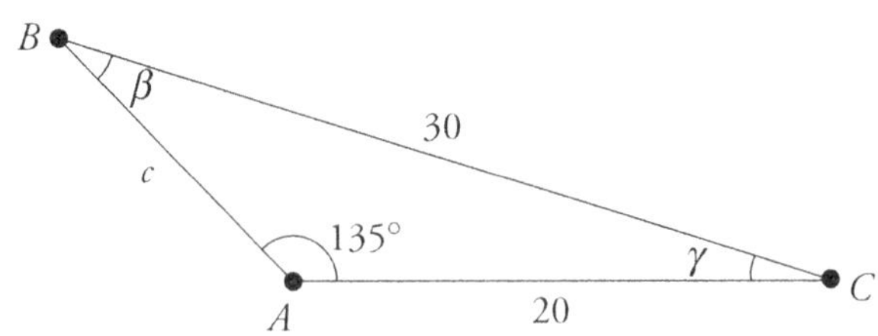

By the Law of Sines Theorem,

$$\frac{a}{\sin\alpha} = \frac{b}{\sin\beta}$$
$$\frac{30}{\sin 135^\circ} = \frac{20}{\sin\beta}$$
$$\sin\beta = \frac{2\sin 135^\circ}{3}$$
$$\sin\beta = \frac{\sqrt{2}}{3}$$

Since $\frac{\sqrt{2}}{3} < 1$, the equation above is a basic sine equation. Since β is the measure of an angle of a triangle, it is in the interval (0°, 180°). Therefore, the above equation has two solutions: $\sin^{-1}\left(\frac{\sqrt{2}}{3}\right)$ and $180^\circ - \sin^{-1}\left(\frac{\sqrt{2}}{3}\right)$.

Case 1: Suppose $\beta = \sin^{-1}\left(\frac{\sqrt{2}}{3}\right)$.

If we round off β to the nearest degree, then $\beta \approx 28^\circ$. Then by *Theorem 6.10*, $\gamma \approx 17^\circ$.

By using the Law of Sines Theorem again,

$$\frac{c}{\sin\gamma} = \frac{a}{\sin\alpha}$$
$$\frac{c}{\sin 17^\circ} \approx \frac{30}{\sin 135^\circ}$$
$$c \approx \frac{30\sin 17^\circ}{\sin 135^\circ} \approx 12.4$$

Therefore, we have one solution to this problem.

$$\alpha = 135^\circ,\ \beta \approx 28^\circ,\ \gamma \approx 17^\circ,\ a = 30,\ b = 20,\ c = 12.4$$

Case 2: Suppose $\beta = 180° - \sin^{-1}\left(\frac{\sqrt{2}}{3}\right)$.

By rounding off β to the nearest degree, we get $\beta = 152°$. However, this is not possible by *Theorem 6.10* because $\alpha + \beta > 180°$. Therefore, this problem has only one solution.

Example 17: Let ΔABC be a triangle so that $a = 11$, $b = 20$, and $\alpha = 35°$.

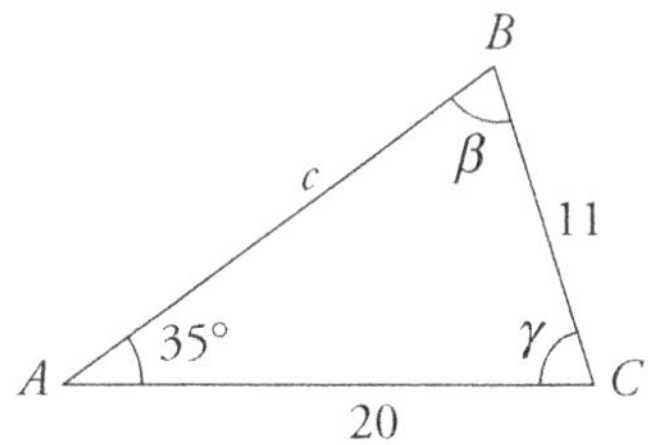

Solve the triangle, if possible.

By the Law of Sines Theorem,

$$\frac{a}{\sin\alpha} = \frac{b}{\sin\beta}$$

$$\frac{11}{\sin 35°} = \frac{20}{\sin\beta}$$

$$\sin\beta = \frac{20\sin 35°}{11}$$

However, $\frac{20\sin 35°}{11} \approx 1.04 > 1$. Therefore, there is no triangle with the given properties.

We can analyze the SSA case geometrically. Suppose a and b are two known numbers and $\angle A$ is an acute angle with measure α. Pick the point C on one side of $\angle A$ so that $AC = b$. Draw a perpendicular from point C to the line containing the other side of $\angle A$. Let the foot of the perpendicular be D. Let h be the length of CD. Now draw a circle with center C and radius a. If this circle intersects line $\overleftrightarrow{AD}$, then it is possible to construct a triangle with a, b, and α known (according to the convention.)

If $h > a$, then the circle with center C and radius a will not intersect line $\overleftrightarrow{AC}$. This is what we observed in Example 17. By *Theorem 6.11*, $h = 20 \sin \alpha$. In Example 16, $h = 20 \sin 35° \approx 11.47$, $a = 11$, and therefore, $h > a$.

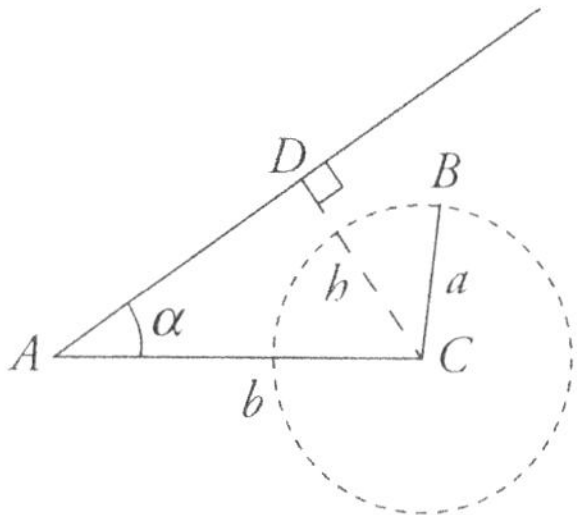

If $a = h$, then ΔABC is a right triangle.

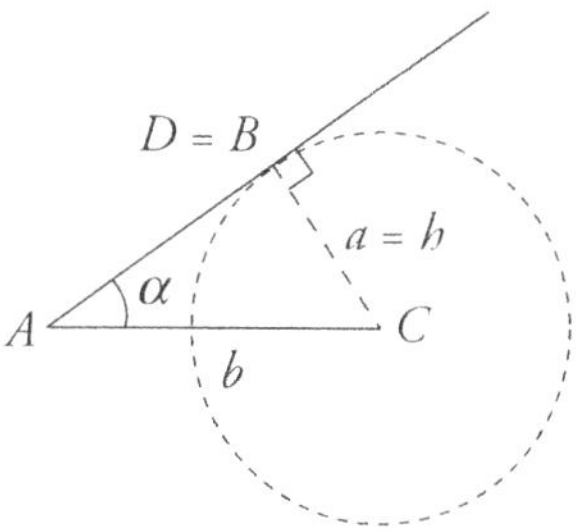

If $b > a > h$, then the circle with center C and radius a intersects the side of α not passing through point C at two points; therefore, there are two solutions. This is what we observed in Example 14.

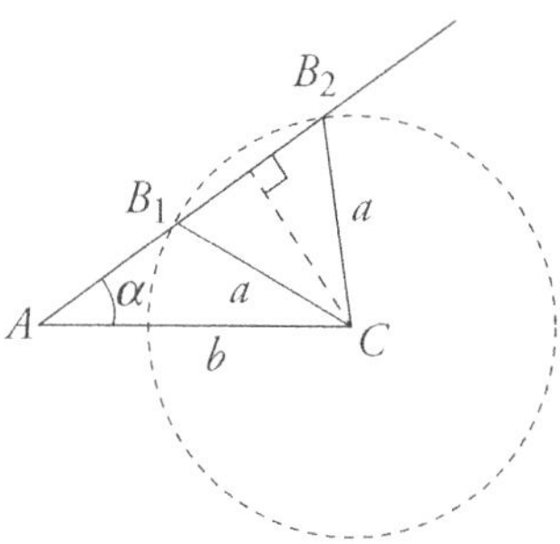

If $a = b$, then the circle with center C and radius a passes through point A and another point on ray $\overrightarrow{AD}$ identified as B in the following figure. In this case, the only solution is an isosceles triangle with $a = b$.

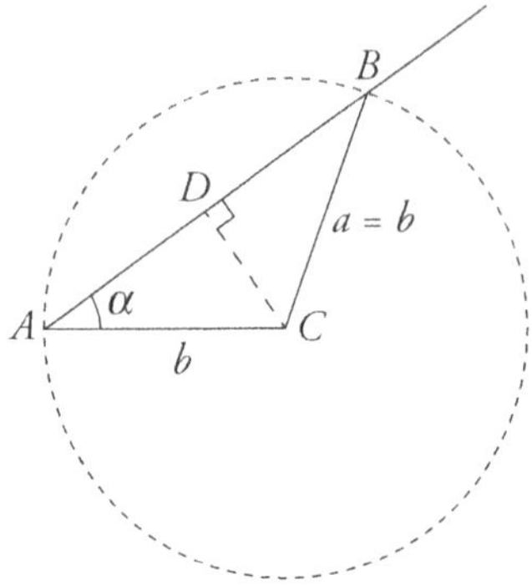

If $a > b$, then the circle with center C and radius a intersects the side of α not passing through point C at only one point. Therefore, there is only one solution. This is what we have observed in Example 15.

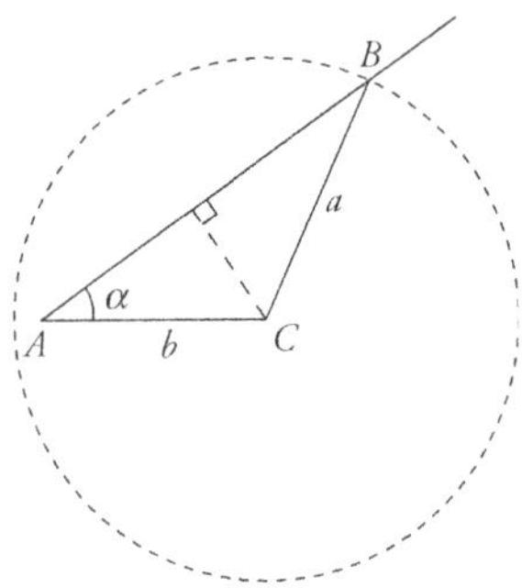

If α is obtuse, then a solution exists only if $a > b$. This is what we have observed in Example 16.

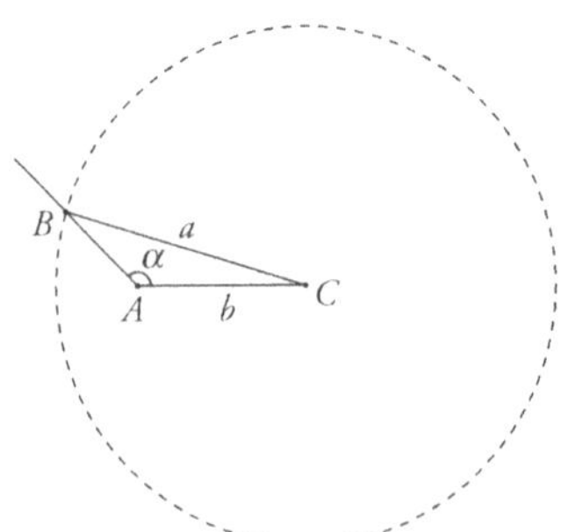

Areas of Triangles

We have the following theorem from geometry. (See *CliffsNotes Geometry Common Core Quick Review.*)

Theorem 6.14: Let ΔABC be a triangle with the accepted convention. Let h_1 be the length of the perpendicular from point A to line $\overleftrightarrow{BC}$, let h_2 be the length of the perpendicular from point B to line $\overleftrightarrow{AC}$, and let h_3 be the length of the perpendicular from point C to line $\overleftrightarrow{AB}$. Then, Area of the triangle $= \frac{1}{2}ah_1 = \frac{1}{2}bh_2 = \frac{1}{2}ch_3$.

We can use *Theorem 6.11* to find the values of h_1, h_2, and h_3. The following is a proof for an acute triangle. Proof for an obtuse triangle is similar.

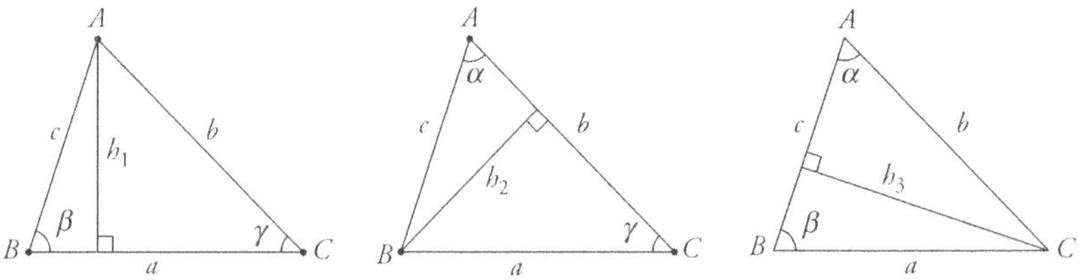

$$h_1 = c \sin\beta = b \sin\gamma$$
$$h_2 = a \sin\gamma = c \sin\alpha$$
$$h_3 = a \sin\beta = b \sin\alpha$$

By substituting the values of h_1, h_2, and h_3 in *Theorem 6.14*, we get the following theorem.

Theorem 6.15: Let ΔABC be a triangle with the adopted convention. Then, the area of the triangle $= \frac{1}{2}ab \sin\gamma = \frac{1}{2}bc \sin\alpha = \frac{1}{2}ca \sin\beta$.

Example 18: Two sides of a triangle have measures of 25 and 12. The measure of the included angle is 51°. Find the area of the triangle.

Let the triangle be ΔABC and let $a = 25$, $b = 12$, and $\gamma = 51°$.

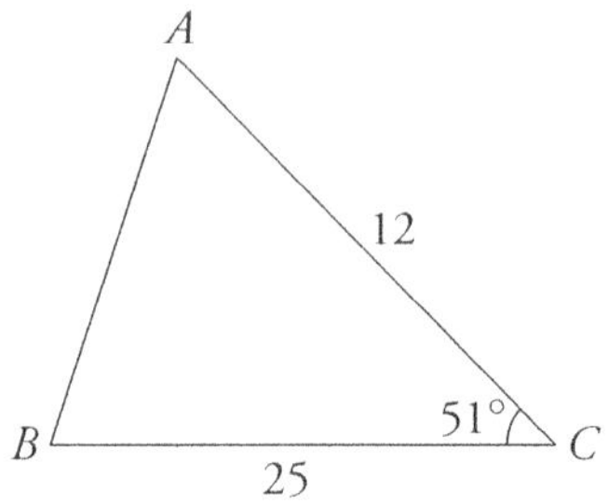

By *Theorem 6.15*,

$$\text{The area of the triangle} = \frac{1}{2}ab\sin\gamma$$
$$= \frac{1}{2}(25)(12)\sin 51°$$
$$\approx 116.57 \text{ square units.}$$

Example 19: Find the area of the triangle shown in the following figure.

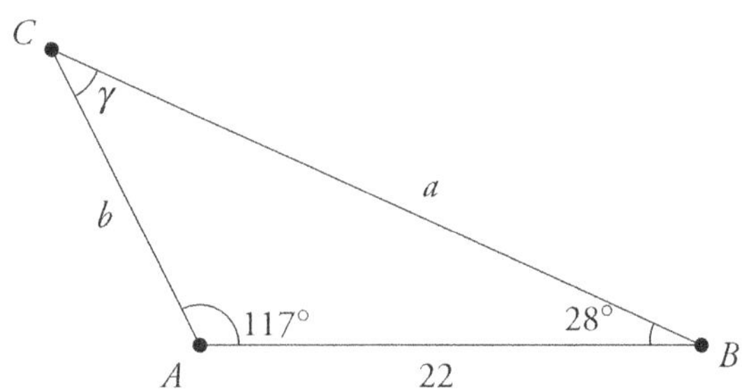

If we can find either a or b, then we can use *Theorem 6.15* to find the area of the triangle. By *Theorem 6.10*, $\gamma = 35°$. By using the Law of Sines Theorem, we can find a.

$$\frac{a}{\sin 117^\circ} = \frac{22}{\sin 35^\circ}$$
$$a = \frac{22 \sin 117^\circ}{\sin 35^\circ}$$

Then by *Theorem 6.15*,

$$\begin{aligned}\text{The area of the triangle} &= \frac{1}{2} ac \sin \beta \\ &= \frac{1}{2}\left(\frac{22 \sin 117^\circ}{\sin 35^\circ}\right)(22) \sin 28^\circ \\ &= 176.49 \text{ square units.}\end{aligned}$$

Example 20: Find the area of an equilateral triangle with a perimeter of 78.

If the perimeter of an equilateral triangle is 78, then the measure of each side is 26. Each of the angles of the triangle is 60°. Therefore, by *Theorem 6.15*,

$$\begin{aligned}\text{The area of the triangle} &= \frac{1}{2} ab \sin \gamma \\ &= \frac{1}{2}(26)(26) \sin 60^\circ \\ &= \frac{1}{2}(26)(26)\frac{\sqrt{3}}{2} \\ &= 169\sqrt{3} \text{ square units.}\end{aligned}$$

The famous Greek philosopher and mathematician Heron (or Hero) developed a formula that calculates the area of a triangle given only the lengths of the three sides. This is known as **Heron's formula.**

If a, b, and c are the lengths of the three sides of a triangle, then the quantity s, $s = \frac{1}{2}(a+b+c)$, is known as the **semi-perimeter** of the triangle.

Theorem 6.16 (Heron's Formula Theorem): If a, b, and c are the lengths of three sides of a triangle, and s is the semi-perimeter, then the area of the triangle is $\sqrt{s(s-a)(s-b)(s-c)}$.

Greeks considered numbers as lengths, the product of two numbers as areas, and the product of three numbers as volumes. By considering a product of four numbers in this theorem, Heron was ahead of his time.

The following is a modern proof of the Heron's Formula Theorem.

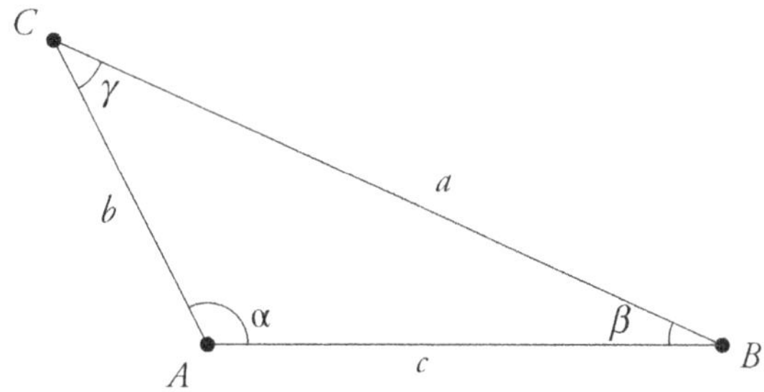

By the Law of Cosines Theorem,

$$a^2 + b^2 - 2ab\cos\gamma = c^2$$
$$a^2 + b^2 - c^2 = 2ab\cos\gamma$$

By squaring both sides of the above equation, we get:

$$\left(a^2 + b^2 - c^2\right)^2 = \left(2ab\cos\gamma\right)^2$$
$$= 4a^2b^2\cos^2\gamma$$

By *Theorem 2.8*, $\cos^2\gamma = 1 - \sin^2\gamma$. Substitute this in the above equation, and we get:

$$\left(a^2 + b^2 - c^2\right)^2 = 4a^2b^2\left(1 - \sin^2\gamma\right)$$
$$= 4a^2b^2 - 4a^2b^2\sin^2\gamma$$
$$= 4a^2b^2 - \left(2ab\sin\gamma\right)^2$$
$$= 4a^2b^2 - \left(4\left(\frac{1}{2}ab\sin\gamma\right)\right)^2$$

By *Theorem 6.15*, $\frac{1}{2}ab\sin\gamma$ is the area of the triangle. Let R be the area of the triangle.

Then $\left(a^2+b^2-c^2\right)^2=4a^2b^2-16R^2$.

Therefore, $16R^2=4a^2b^2-\left(a^2+b^2-c^2\right)^2$.

By using the Difference of Squares Identity,

$$\begin{aligned}16R^2&=4a^2b^2-\left(a^2+b^2-c^2\right)^2\\&=\left[2ab+\left(a^2+b^2-c^2\right)\right]\left[2ab-\left(a^2+b^2-c^2\right)\right]\\&=\left[\left(a^2+2ab+b^2\right)-c^2\right]\left[c^2-\left(a^2-2ab+b^2\right)\right]\\&=\left[(a+b)^2-c^2\right]\left[c^2-(a-b)^2\right]\end{aligned}$$

By using the Difference of Squares Identity again, we get:

$$\begin{aligned}16R^2&=\left[(a+b)^2-c^2\right]\left[c^2-(a-b)^2\right]\\&=\left[((a+b)+c)((a+b)-c)\right]\left[(c+(a-b))(c-(a-b))\right]\\&=\left[(a+b+c)((a+b+c)-2c)\right]\left[((a+b+c)-2b)((a+b+c)-2a)\right]\end{aligned}$$

Now, $(a + b + c) = 2s$, by the definition of semi-perimeter. Substitute $2s$ for every $(a + b + c)$ in the above equation:

$$\begin{aligned}16R^2&=\left[(a+b+c)((a+b+c)-2c)\right]\left[((a+b+c)-2b)((a+b+c)-2a)\right]\\&=\left[(2s)(2s-2c)\right]\left[(2s-2b)(2s-2a)\right]\\&-16\left[s(s-c)(s-b)(s-a)\right]\end{aligned}$$

Divide both sides of the above equation by 16 and then use the Square-Root Principle Theorem,

$$R=\pm\sqrt{s(s-a)(s-b)(s-c)}$$

Since R is a positive number (area of the triangle),

$$R=\sqrt{s(s-a)(s-b)(s-c)}$$

Example 21: Find the area of a triangle if its sides measure 31, 44, and 60.

The semi-perimeter is $s = \dfrac{31+44+60}{2} = \dfrac{135}{2}$.

Then $s-31 = \dfrac{135}{2} - 31 = \dfrac{73}{2}$, $s-44 = \dfrac{135}{2} - 44 = \dfrac{47}{2}$, and $s-60 = \dfrac{135}{2} - 60 = \dfrac{15}{2}$.

Let the area of the triangle be *R*. Then by using the Heron's Formula Theorem:

$$\begin{aligned} R &= \sqrt{s(s-a)(s-b)(s-c)} \\ &= \sqrt{\left(\frac{135}{2}\right)\left(\frac{73}{2}\right)\left(\frac{47}{2}\right)\left(\frac{15}{2}\right)} \\ &= \frac{1}{4}\sqrt{(135)(73)(47)(15)} \\ &= \frac{1}{4}\sqrt{6{,}947{,}775} \\ &\approx \frac{1}{4}(2635.86) \\ &\approx 658.97 \text{ square units} \end{aligned}$$

Chapter Check-Out

Questions

1. In a right triangle, the two legs have lengths of 8 inches and 10 inches. Find the measure of the angle opposite the 10-inch side.
2. In a right triangle, the side opposite the 38° angle is 6.4 inches long. Find the length of the hypotenuse.
3. You are standing 400 feet from the base of a building. The angle of elevation to the top of the building is 28°. Find the height of the building.
4. Find the measure of the smallest angle in a triangle if the three sides measure 7 inches, 8 inches, and 10 inches.

5. Two sides of a triangle measure 12 feet and 18 feet, and the angle between them measures 16°. Find the length of the third side.
6. True or False: In some cases, the Law of Sines Theorem will not provide you with a unique answer.
7. Find the area of a triangle if its sides measure 10 feet, 14 feet, and 20 feet.
8. Find the area of a triangle if two of its sides measure 8 inches and 14 inches, and the angle between them measures 72°.

Answers

1. 51.34°
2. 10.395 inches
3. 213 feet
4. 44.05°
5. 7.26 feet
6. True
7. 64.99 square feet
8. 53.26 square inches

Chapter 7

TRIGONOMETRIC IDENTITIES AND EQUATIONS

Chapter Check-In

- ❑ Defining several fundamental trigonometric identities
- ❑ Using sum and difference formulas to extend the fundamental identities
- ❑ Using double- and half-angle identities
- ❑ Understanding the tangent identities
- ❑ Using the product-to-sum and sum-to-product identities

Common Core Standard: Prove and apply trigonometric identities

Prove the Pythagorean identity $\sin^2\theta + \cos^2\theta = 1$ and use it to find sin (θ), $\cos(\theta)$, or $\tan(\theta)$ given $\sin(\theta)$, $\cos(\theta)$, or $\tan(\theta)$ and the quadrant of the angle. (HSF.TF.C.8)

Prove the addition and subtraction formulas for sine, cosine, and tangent and use them to solve problems. (HSF.TF.C.9)

Use inverse functions to solve trigonometric equations that arise in modeling contexts; evaluate the solutions using technology, and interpret them in terms of the context. (HSF.TF.B.7)

Please also refer to the standards for mathematical practice, found here: www.corestandards.org/Math/Practice/.

Some equations are true for only a few or countably many numbers. Some equations are true for all real numbers or almost all real numbers. This second type of equation is called an *identity.* The knowledge of these identities is useful in solving more complex equations. As you explore new properties of trigonometric functions, new identities are established, and these new identities can be used to establish still more identities, and so on.

Trigonometry Theorems

The trigonometry theorems in this book are numbered for organizational purposes. For example, *Theorem 2.2* is the second theorem given in Chapter 2. Study these theorems by their content, not by their number, as the theorem numbering has no significance outside of this book.

Fundamental Identities

We have already seen some trigonometric identities. For example, consider the definition of the tangent function: $\frac{\sin\theta}{\cos\theta} = \tan\theta$. It is true for all numbers θ, except where $\cos\theta = 0$.

There are more numbers where the tangent function is defined rather than undefined. (See the graph of the tangent function in Chapter 3, p. 66.) Therefore, we say the definition of the tangent function is a basic trigonometric identity.

We have seen the following fundamental (basic) trigonometric identities so far—the definitions of tangent, cosecant, secant, cotangent numbers (see Chapter 2):

$$\tan\theta = \frac{\sin\theta}{\cos\theta}, \quad \csc\theta = \frac{1}{\sin\theta}, \quad \sec\theta = \frac{1}{\cos\theta}, \quad \cot\theta = \frac{1}{\tan\theta}$$

From *Theorem 2.19*, we have $\cot = \frac{\cos\theta}{\sin\theta}$.

From *Theorem 2.10*, we have $\cos(-\theta) = \cos\theta$ and $\sin(-\theta) = -\sin\theta$.

From *Theorem 2.9*, we have $\sin(2n\pi + \theta) = \sin\theta$ and $\cos(2n\pi + \theta) = \cos\theta$.

From *Theorem 2.11*, we have $\sin(\pi - \theta) = \sin\theta$ and $\cos(\pi - \theta) = -\cos\theta$.

From *Theorem 2.8*, we have $\cos^2\theta + \sin^2\theta = 1$.

The following theorem follows directly from *Theorem 2.8*, but the result is by itself a useful identity.

Theorem 7.1: $1 + \tan^2 \theta = \sec^2 \theta$.

Proof: If you divide both sides of $\cos^2 \theta + \sin^2 \theta = 1$ by $\cos^2 \theta$, clearly, $\cos \theta$ has to be non-zero. We usually do not mention this requirement since the theorem is true for many more numbers than the numbers where $\cos \theta \neq 0$. We will adopt this practice for identities.

The following two theorems also follow directly from *Theorem 2.8*.

Theorem 7.2: $\sin^2 \theta = 1 - \cos^2 \theta$.

Theorem 7.3: $\cos^2 \theta = 1 - \sin^2 \theta$.

Sometimes you may encounter an equation that looks like an identity, but you are not sure. That means you do not know if the left side of the equation is equal to the right side of the equation for almost all numbers. Finding out if this is so is called *verifying an identity.*

To verify a given equation as an identity, you start from one side of the equation and perform valid arithmetic or algebraic operations, or use known trigonometric identities to simplify that side until you get it to equal the other side of the equation. Another technique is to simplify one side of the equation as much as you can, and then simplify the other side of the equation as much as you can; if the two results are equal, then the equation is an identity.

Example 1: Verify the identity $\cos \theta + \sin \theta \tan \theta = \sec \theta$.

Let's start with the left side: $\cos \theta + \sin \theta \tan \theta$.

There is no visible arithmetic or algebra to do. Therefore, we will use the trigonometric identity $\tan\theta = \dfrac{\sin\theta}{\cos\theta}$.

$$\text{left side} = \cos\theta + \sin\theta\left(\frac{\sin\theta}{\cos\theta}\right)$$

Now, we see that we can do some arithmetic.

$$\text{left side} = \cos\theta + \frac{\sin^2\theta}{\cos\theta}$$

We can do more arithmetic: addition of fractions.

$$\text{left side} = \frac{\cos^2\theta + \sin^2\theta}{\cos\theta}$$

At this point, you should realize that you can use the trigonometric identity: $\cos^2\theta + \sin^2\theta = 1$.

$$\text{left side} = \frac{1}{\cos\theta}$$

Now, we can use the definition of the secant function: $\sec\theta = \dfrac{1}{\cos\theta}$.

$$\text{left side} = \sec\theta$$

We have verified that the left side is equal to the right side. The following is the proof of Example 1.

$$\begin{aligned}
\text{left side} &= \cos\theta + \sin\theta\tan\theta \\
&= \cos\theta + \sin\theta\left(\frac{\sin\theta}{\cos\theta}\right) \\
&= \cos\theta + \frac{\sin^2\theta}{\cos\theta} \\
&= \frac{\cos^2\theta + \sin^2\theta}{\cos\theta} \\
&= \frac{1}{\cos\theta} \\
&= \sec\theta \\
&= \text{right side}
\end{aligned}$$

In Example 2 that follows, we have $\sec\theta + 1$ in the numerator of the right side. Therefore, we choose to multiply both the numerator and the denominator of the left side by $\sec\theta + 1$ after selecting the left side. In the given solution, the rest of the details of the thinking process are omitted.

Example 2: Verify the identity $\frac{\tan\theta}{\sec\theta-1}=\frac{\sec\theta+1}{\tan\theta}$.

$$\begin{aligned}
\text{left side} &= \frac{\tan\theta}{\sec\theta-1}\\
&= \left(\frac{\tan\theta}{\sec\theta-1}\right)\left(\frac{\sec\theta+1}{\sec\theta+1}\right)\\
&= \frac{\tan\theta(\sec\theta+1)}{\sec^2\theta-1}\\
&= \frac{\tan\theta(\sec\theta+1)}{(1+\tan^2\theta)-1}\\
&= \frac{\tan\theta(\sec\theta+1)}{\tan^2\theta}\\
&= \frac{\sec\theta+1}{\tan\theta}\\
&= \text{right side}
\end{aligned}$$

The following is an indirect way of proving the previous identity. The problem is that initially, we do not know that $\frac{\tan\theta}{\sec\theta-1}$ is equal to $\frac{\sec\theta+1}{\tan\theta}$. If we did, then there would be nothing to prove. However, we can start with an identity that we know is true: $1 + \tan^2\theta = \sec^2\theta$.

$$\begin{aligned}
1+\tan^2\theta &= \sec^2\theta\\
\tan^2\theta &= \sec^2\theta-1\\
\tan^2\theta &= (\sec\theta-1)(\sec\theta+1)
\end{aligned}$$

In the last step, we used the all-important difference of squares identity. Now, divide both sides by $\tan\theta(\sec\theta-1)$, and we get $\frac{\tan\theta}{\sec\theta-1}=\frac{\sec\theta+1}{\tan\theta}$ as desired.

Example 3: Verify the identity $\dfrac{\sin^2\theta+2\cos\theta-1}{\sin^2\theta+3\cos\theta-3}=\dfrac{1}{1-\sec\theta}$.

$$\begin{aligned}
\text{left side} &= \frac{\sin^2\theta+2\cos\theta-1}{\sin^2\theta+3\cos\theta-3}\\
&= \frac{1-\cos^2\theta+2\cos\theta-1}{1-\cos^2\theta+3\cos\theta-3}\\
&= \frac{-\left(\cos^2\theta-2\cos\theta\right)}{-\left(\cos^2\theta-3\cos\theta+2\right)}\\
&= \frac{\cos\theta\left(\cos\theta-2\right)}{\left(\cos\theta-2\right)\left(\cos\theta-1\right)}\\
&= \frac{\cos\theta}{\cos\theta-1}\\
&= \frac{\dfrac{\cos\theta}{\cos\theta}}{\dfrac{\cos\theta-1}{\cos\theta}}\\
&= \frac{1}{1-\dfrac{1}{\cos\theta}}\\
&= \frac{1}{1-\sec\theta}\\
&= \text{right side}
\end{aligned}$$

This verifies that the given equation is an identity.

Sum and Difference Identities

Perhaps the most important trigonometric identities are those known as the sum and difference identities. But before we can get those identities, we need to establish one more theorem.

Theorem 7.4: $\sin\left(\frac{\pi}{2}+\theta\right)=\cos\theta$ and $\cos\left(\frac{\pi}{2}+\theta\right)=-\sin\theta$.

Proof: We will use the unit circle to prove this theorem. It is possible that θ is in one of the quadrants or θ is a quadrant angle. First, suppose that θ is not a quadrant angle; then θ is in one of the four quadrants. The following figures show both θ and $\frac{\pi}{2}+\theta$.

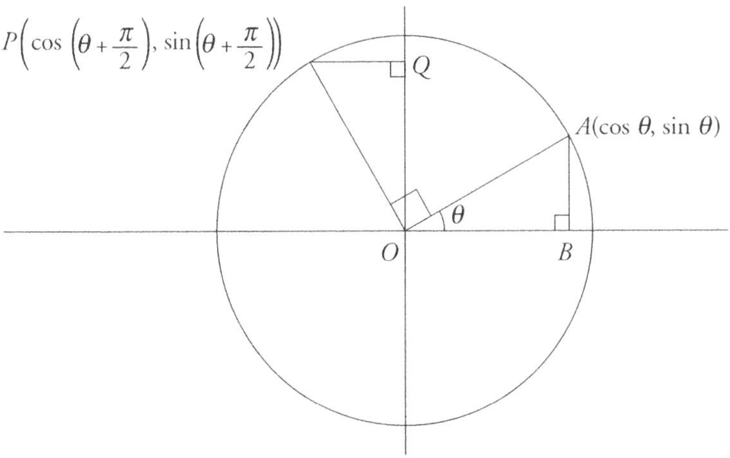

(a)

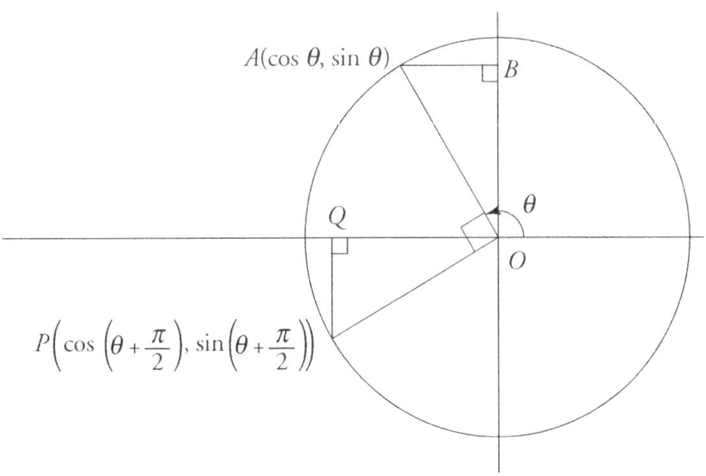

(b)

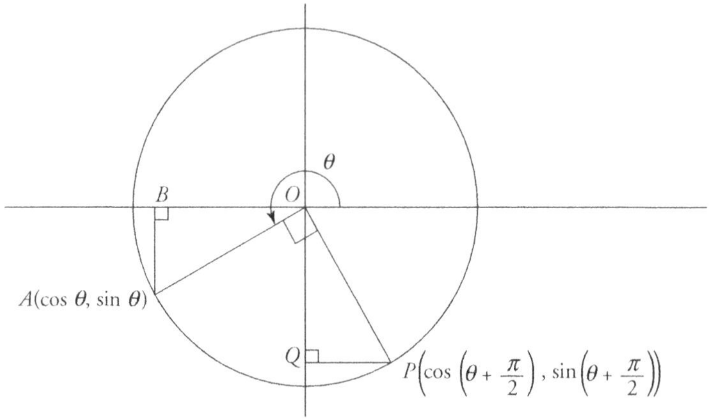

(c)

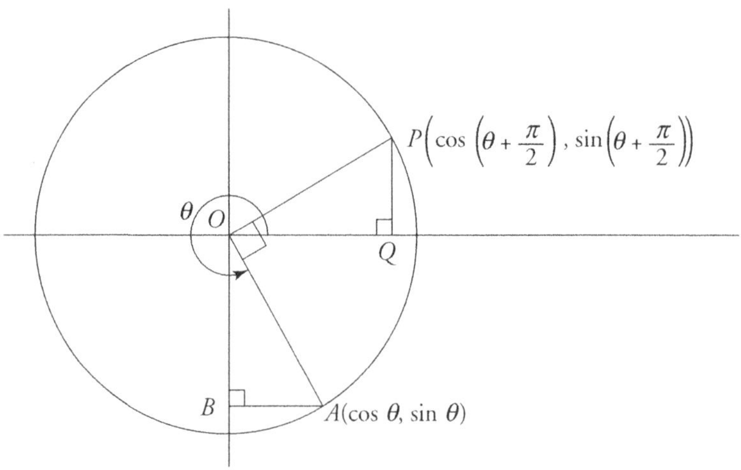

(d)

In figure (a), θ is in the first quadrant; in figure (b), θ is in the second quadrant; in figure (c), θ is in the third quadrant; and in figure (d), θ is in the fourth quadrant.

If θ is in the first quadrant, then from figure (a), $|\angle AOB| = 90° - |\angle QOA|$ and $|\angle POQ| = 90° - |\angle QOA|$.

Therefore, $|\angle AOB| = |\angle POQ|$, $\angle OBA$ and $\angle OQP$ are right angles, and $OA = OP$. Therefore, by the SAA Theorem, ΔOAB and ΔOPQ are congruent. Therefore, $PQ = AB$ and $OQ = OB$. That is, $-\cos\left(\frac{\pi}{2}+\theta\right)=\sin\theta$ and $\sin\left(\frac{\pi}{2}+\theta\right)=\cos\theta$.

If θ is in the second quadrant, then from figure (b), ΔOAB and ΔOPQ are congruent, and $PQ = AB$ and $OQ = OB$. That is, $-\sin\left(\frac{\pi}{2}+\theta\right)=-\cos\theta$ and $-\cos\left(\frac{\pi}{2}+\theta\right)=\sin\theta$.

If θ is in the third quadrant, then from figure (c), ΔOAB and ΔOPQ are congruent, and $PQ = AB$ and $OQ = OB$. That is, $\cos\left(\frac{\pi}{2}+\theta\right)=-\sin\theta$ and $-\sin\left(\frac{\pi}{2}+\theta\right)=-\cos\theta$.

If θ is in the fourth quadrant, then from figure (d), ΔOAB and ΔOPQ are congruent, and $PQ = AB$ and $OQ = OB$. That is, $\sin\left(\frac{\pi}{2}+\theta\right)=\cos\theta$ and $\cos\left(\frac{\pi}{2}+\theta\right)=-\sin\theta$.

Now suppose θ is a quadrant angle. If θ is 0 or a multiple of 2π, then $\cos\theta = 1$, $\sin\theta = 0$, $\cos\left(\frac{\pi}{2}+\theta\right)=0$, and $\sin\left(\frac{\pi}{2}+\theta\right)=1$. So, $\sin\left(\frac{\pi}{2}+\theta\right)=\cos\theta$ and $\cos\left(\frac{\pi}{2}+\theta\right)=-\sin\theta$ are still true.

If θ is $\frac{\pi}{2}$, or $-\frac{3\pi}{2}$, or $2k\pi+\frac{\pi}{2}$, or $-2k\pi-\frac{3\pi}{2}$, where k is a positive integer, then $\cos\theta = 0$, $\sin\theta = 1$, $\cos\left(\frac{\pi}{2}+\theta\right)=-1$, and $\sin\left(\frac{\pi}{2}+\theta\right)=0$. The theorem is true in this case.

If θ is π, or $-\pi$, or $2k\pi+\pi$, or $-2k\pi-\pi$, where k is a positive integer, then $\cos\theta = -1$, $\sin\theta = 0$, $\cos\left(\frac{\pi}{2}+\theta\right)=0$, and $\sin\left(\frac{\pi}{2}+\theta\right)=-1$. The theorem is also true in this case.

If θ is $\frac{3\pi}{2}$, or $-\frac{\pi}{2}$, or $2k\pi+\frac{3\pi}{2}$, or $-2k\pi-\frac{\pi}{2}$, where k is a positive integer, then $\cos\theta = 0$, $\sin\theta = -1$, $\cos\left(\frac{\pi}{2}=\theta\right)=1$, and $\sin\left(\frac{\pi}{2}+\theta\right)=0$.

The theorem is true in this case as well.

With *Theorem 7.4* in place, we are ready to prove the first sum and difference identity. The following is a theorem from geometry. (See *CliffsNotes Geometry Common Core Quick Review.*)

Theorem 7.5: In a circle, if the central angles subtended by two arcs are equal in measure, then their corresponding chords are equal in measure.

Theorem 7.6: $\cos(\alpha - \beta) = \cos\alpha\cos\beta + \sin\alpha\sin\beta$.

Proof: For *Theorem 7.6*, if $\alpha = 0$ or $\beta = 0$, then the theorem is easy to prove. For example, assume that $\alpha = 0$. Then the left side of the identity is $\cos(0 - \beta) = \cos(-\beta) = \cos\beta$, by *Theorem 2.10*. The right side of the identity is $\cos 0 \cos\beta + \sin 0 \sin\beta = \cos\beta$, since $\cos 0 = 1$ and $\sin 0 = 0$. Therefore, the theorem is true if at least one of the numbers is 0.

It is also easy to prove that *Theorem 7.6* is true if at least one of the angles is 2π. For example, assume $\beta = 2\pi$. Then the left side of the identity is $\cos(\alpha - 2\pi) = \cos(-(2\pi - \alpha)) = \cos(2\pi - \alpha) = \cos(2\pi + (-\alpha)) = \cos(-\alpha)$.

The last equality follows from *Theorem 2.9*. Since $\cos(-\alpha) = \cos\alpha$, the left side is reduced to $\cos\alpha$. The right side of the identity is $\cos\alpha\cos 2\pi + \sin\alpha\sin 2\pi = \cos\alpha$, since $\cos 2\pi = 1$ and $\sin 2\pi = 0$. Therefore, the theorem is true if at least one of the numbers is 2π.

Let us assume that $0 < \beta < \alpha < 2\pi$. We will use the unit circle and consider α and β as angles in radian measure. The following figures are just a guide to the argument.

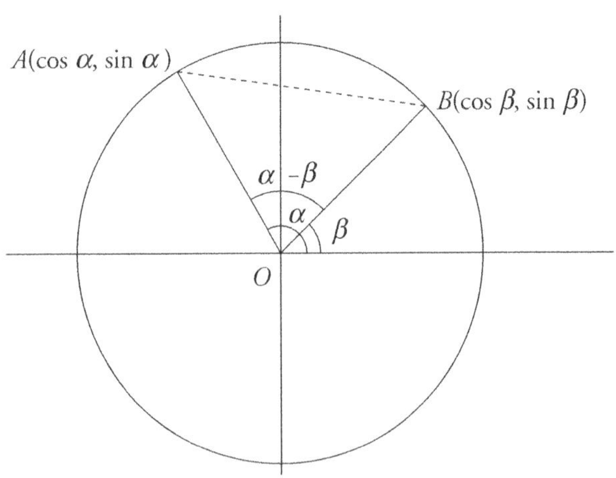

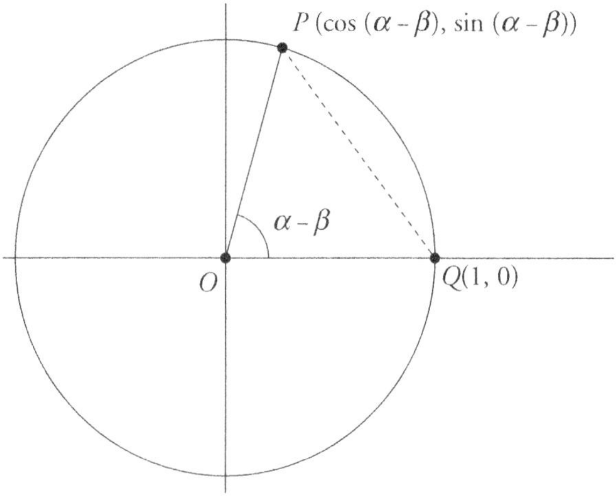

Since $\alpha - \beta = (\alpha - \beta) - 0$, the central angles $\angle AOB$ and $\angle POQ$ are equal. Then by *Theorem 7.5*, $AB = PQ$. By the two-dimensional distance formula,

$$\begin{aligned} AB^2 &= (\cos\alpha - \cos\beta)^2 + (\sin\alpha - \sin\beta)^2 \\ &= \cos^2\alpha - 2\cos\alpha\cos\beta + \cos^2\beta + \sin^2\alpha - 2\sin\alpha\sin\beta + \sin^2\beta \\ &= (\cos^2\alpha + \sin^2\alpha) + (\cos^2\beta + \sin^2\beta) - 2(\cos\alpha\cos\beta + \sin\alpha\sin\beta) \\ &= 2 - 2(\cos\alpha\cos\beta + \sin\alpha\sin\beta) \end{aligned}$$

and

$$\begin{aligned} PQ^2 &= (\cos(\alpha-\beta) - 1)^2 + (\sin(\alpha-\beta) - 0)^2 \\ &= \cos^2(\alpha-\beta) - 2\cos(\alpha-\beta) + 1 + \sin^2(\alpha-\beta) \\ &= 1 + (\cos^2(\alpha-\beta) + \sin^2(\alpha-\beta)) - 2\cos(\alpha-\beta) \\ &= 2 - 2\cos(\alpha-\beta) \end{aligned}$$

Since $AB = PQ$, $AB^2 = PQ^2$, and $2 - 2(\cos\alpha\cos\beta + \sin\alpha\sin\beta) = 2 - 2\cos(\alpha - \beta)$.

Subtract 2 from both sides and then divide both sides by –2 and you get

$$\cos(\alpha - \beta) = \cos\alpha\cos\beta + \sin\alpha\sin\beta$$

Now, suppose at least one of α or β is more than 2π. Let us say $\alpha > 2\pi$. Then $\alpha = 2n\pi + \theta$ for some number $0 < \theta < 2\pi$ and for some positive integer n.

Then

$$\begin{aligned}\cos(\alpha-\beta) &= \cos(2n\pi+\theta-\beta)\\ &= \cos(2n\pi+(\theta-\beta))\\ &= \cos(\theta-\beta)\\ &= \cos\theta\cos\beta+\sin\theta\sin\beta, \text{ by } \textit{Theorem 2.9} \text{ and the previous case}\end{aligned}$$

However, $\cos\theta = \cos(2n\pi + \theta) = \cos\alpha$ and $\sin\theta = \sin(2n\pi + \theta) = \sin\alpha$, again by *Theorem 2.9*. Therefore, $\cos(\alpha - \beta) = \cos\alpha\cos\beta + \sin\alpha\sin\beta$. Proof of the case, if at least one of the numbers is less than -2π, is similar.

Theorem 7.7: $\cos(\alpha + \beta) = \cos\alpha\cos\beta - \sin\alpha\sin\beta$.

Proof:

$$\cos(\alpha+\beta) = \cos(\alpha-(-\beta))$$

Now we can use *Theorem 7.6*. Therefore,

$$\begin{aligned}\cos(\alpha+\beta) &= \cos\alpha\cos(-\beta)+\sin\alpha\sin(-\beta)\\ &= \cos\alpha\cos\beta-\sin\alpha\sin\beta\end{aligned}$$

The last equality follows from *Theorem 2.10*.

Theorem 7.8: $\sin(\alpha + \beta) = \sin\alpha\cos\beta + \cos\alpha\sin\beta$.

Proof: By *Theorem 7.4*, $\cos\left(\frac{\pi}{2}+(\alpha+\beta)\right) = -\sin(\alpha+\beta)$. Therefore,

$$\begin{aligned}\sin(\alpha+\beta) &= -\cos\left(\frac{\pi}{2}+(\alpha+\beta)\right)\\ &= -\cos\left(\left(\frac{\pi}{2}+\alpha\right)+\beta\right)\end{aligned}$$

Then by *Theorem 7.7*,

$$\begin{aligned}\sin(\alpha+\beta) &= -\left[\cos\left(\frac{\pi}{2}+\alpha\right)\cos\beta - \sin\left(\frac{\pi}{2}+\alpha\right)\sin\beta\right] \\ &= -[-\sin\alpha\cos\beta - \cos\alpha\sin\beta] \\ &= \sin\alpha\cos\beta + \cos\alpha\sin\beta\end{aligned}$$

The second-to-last equality above follows from *Theorem 7.4.*

Theorem 7.9: $\sin(\alpha - \beta) = \sin\alpha\cos\beta - \cos\alpha\sin\beta$.

Proof:

$$\begin{aligned}\sin(\alpha-\beta) &= \sin(\alpha+(-\beta)) \\ &= \sin\alpha\cos(-\beta) + \cos\alpha\sin(-\beta), \text{by } \textit{Theorem 7.8} \\ &= \sin\alpha\cos\beta - \cos\alpha\sin\beta\end{aligned}$$

The last equality above follows from *Theorem 2.10.*

The identities in *Theorems 7.6–7.9* are known as sum and difference identities:

$$\begin{aligned}\cos(\alpha-\beta) &= \cos\alpha\cos\beta + \sin\alpha\sin\beta \\ \cos(\alpha+\beta) &= \cos\alpha\cos\beta - \sin\alpha\sin\beta \\ \sin(\alpha+\beta) &= \sin\alpha\cos\beta + \cos\alpha\sin\beta \\ \sin(\alpha-\beta) &= \sin\alpha\cos\beta - \cos\alpha\sin\beta\end{aligned}$$

Example 4: Find the exact value of cos 15°.

Since 15° = 45° – 30°,

$$\begin{aligned}\cos 15^\circ &= \cos(45^\circ - 30^\circ) \\ &= \cos 45^\circ \cos 30^\circ + \sin 45^\circ \sin 30^\circ, \text{ by } \textit{Theorem 7.6} \\ &= \left(\frac{\sqrt{2}}{2}\right)\left(\frac{\sqrt{3}}{2}\right) + \left(\frac{\sqrt{2}}{2}\right)\left(\frac{1}{2}\right) \\ &= \frac{\sqrt{2}\left(\sqrt{3}+1\right)}{4}\end{aligned}$$

Example 5: Find $\sin(\alpha + \beta)$ if $\sin\alpha = -\frac{4}{5}$, $\cos\beta = \frac{15}{17}$, and α and β are fourth quadrant angles.

First, find $\cos\alpha$ and $\sin\beta$. The sine is negative and the cosine is positive in the fourth quadrant. Using *Theorems 2.8* and *7.8*,

$$\sin^2\alpha + \cos^2\alpha = 1 \qquad \sin^2\beta + \cos^2\beta = 1$$

$$\left(\frac{-4}{5}\right)^2 + \cos^2\alpha = 1 \qquad \sin^2\beta + \left(\frac{15}{17}\right)^2 = 1$$

$$\cos^2 = 1 - \frac{16}{25} \qquad \sin^2\beta = 1 - \frac{225}{289}$$

$$\cos^2\alpha = \frac{9}{25} \qquad \sin^2\beta = \frac{64}{289}$$

$$\cos\alpha = \frac{3}{5} \qquad \sin\beta = -\frac{8}{17}$$

$$\sin(\alpha + \beta) = \left(-\frac{4}{5}\right)\left(\frac{15}{17}\right) + \left(\frac{3}{5}\right)\left(-\frac{8}{17}\right)$$

$$\sin(\alpha + \beta) = \left(-\frac{60}{85}\right) + \left(-\frac{24}{85}\right)$$

$$\sin(\alpha + \beta) = -\frac{84}{85}$$

Example 6: Show that $\sin\left(\frac{\pi}{2} - \theta\right) = \cos\theta$.

By *Theorem 7.9*,

$$\begin{aligned}\sin\left(\frac{\pi}{2} - \theta\right) &= \sin\frac{\pi}{2}\cos\theta - \cos\frac{\pi}{2}\sin\theta \\ &= (1)\cos\theta - (0)\sin\theta \\ &= \cos\theta\end{aligned}$$

Example 7: Show that $\cos\left(\frac{\pi}{2} - \theta\right) = \sin\theta$.

By *Theorem 7.6*,

$$\begin{aligned}\cos\left(\frac{\pi}{2} - \theta\right) &= \cos\frac{\pi}{2}\cos\theta + \sin\frac{\pi}{2}\sin\theta \\ &= \sin\theta\end{aligned}$$

Theorem 7.10: $\tan(\alpha+\beta)=\dfrac{\tan\alpha+\tan\beta}{1-\tan\alpha\tan\beta}$.

Proof: By definition, $\tan(\alpha+\beta)=\dfrac{\sin(\alpha+\beta)}{\cos(\alpha+\beta)}$. Then, by *Theorems 7.7* and *7.8*,

$$\tan(\alpha+\beta)=\frac{\sin(\alpha+\beta)}{\cos(\alpha+\beta)}$$

$$=\frac{\sin\alpha\cos\beta+\cos\alpha\sin\beta}{\cos\alpha\cos\beta-\sin\alpha\sin\beta}$$

$$=\frac{\dfrac{\sin\alpha\cos\beta}{\cos\alpha\cos\beta}+\dfrac{\cos\alpha\sin\beta}{\cos\alpha\cos\beta}}{\dfrac{\cos\alpha\cos\beta}{\cos\alpha\cos\beta}-\dfrac{\sin\alpha\sin\beta}{\cos\alpha\cos\beta}}$$

$$=\frac{\tan\alpha+\tan\beta}{1-\tan\alpha\tan\beta}$$

Theorem 7.11: $\tan(\alpha-\beta)=\dfrac{\tan\alpha-\tan\beta}{1+\tan\alpha\tan\beta}$.

Proof: By definition, $\tan(\alpha-\beta)=\dfrac{\sin(\alpha-\beta)}{\cos(\alpha-\beta)}$. Using *Theorems 7.6* and *7.9*,

$$\tan(\alpha-\beta)=\frac{\sin(\alpha-\beta)}{\cos(\alpha-\beta)}$$

$$=\frac{\sin\alpha\cos\beta-\cos\alpha\sin\beta}{\cos\alpha\cos\beta+\sin\alpha\sin\beta}$$

$$=\frac{\dfrac{\sin\alpha\cos\beta}{\cos\alpha\cos\beta}-\dfrac{\cos\alpha\sin\beta}{\cos\alpha\cos\beta}}{\dfrac{\cos\alpha\cos\beta}{\cos\alpha\cos\beta}+\dfrac{\sin\alpha\sin\beta}{\cos\alpha\cos\beta}}$$

$$=\frac{\tan\alpha-\tan\beta}{1+\tan\alpha\tan\beta}$$

Example 8: Find the exact value of tan 75°.

Because 75° = 45° + 30°, tan 75° = tan (45° + 30°). Then by *Theorem 7.10,*

$$\begin{aligned}\tan 75^\circ &= \frac{\tan 45^\circ + \tan 30^\circ}{1 - \tan 45^\circ \tan 30^\circ}\\ &= \frac{1 + \frac{1}{\sqrt{3}}}{1 - (1)\left(\frac{1}{\sqrt{3}}\right)}\\ &= \frac{\left(\frac{\sqrt{3}+1}{\sqrt{3}}\right)\left(\sqrt{3}\right)}{\left(\frac{\sqrt{3}-1}{\sqrt{3}}\right)\sqrt{3}}\\ &= \frac{\sqrt{3}+1}{\sqrt{3}-1}\\ &= \left(\frac{\sqrt{3}+1}{\sqrt{3}-1}\right)\left(\frac{\sqrt{3}+1}{\sqrt{3}+1}\right)\\ &= \frac{3 + 2\sqrt{3} + 1}{3 - 1}\\ &= 2 + \sqrt{3}\end{aligned}$$

Example 9: Verify that tan (180° – x) = –tan x, for all x where tan x is defined.

By *Theorem 7.11,*

$$\begin{aligned}\tan(180^\circ - x) &= \frac{\tan 180^\circ - \tan x}{1 + \tan 180^\circ \tan x}\\ &= \frac{0 - \tan x}{1 + (0)\tan x}\\ &= -\tan x\end{aligned}$$

Example 10: Verify that tan (180° + x) = tan x, for all x where tan x is defined.

By *Theorem 7.10*,

$$\begin{aligned}\tan(180^\circ + x) &= \frac{\tan 180^\circ + \tan x}{1 - \tan 180^\circ \tan x}\\ &= \frac{0 + \tan x}{1-(0)\tan x}\\ &= \tan x\end{aligned}$$

Double-Angle Identities

Special cases of the sum and difference formulas for sine and cosine yield what are known as the *double-angle identities* and the *half-angle identities*. These, too, are very important identities.

Theorem 7.12: $\sin 2\theta = 2\sin\theta\cos\theta$.

Proof: By *Theorem 7.8*,

$$\begin{aligned}\sin 2\theta &= \sin(\theta+\theta)\\ &= \sin\theta\cos\theta + \cos\theta\sin\theta\\ &= 2\sin\theta\cos\theta\end{aligned}$$

Theorem 7.13: $\cos 2\theta = \cos^2\theta - \sin^2\theta$.

Proof: By *Theorem 7.7*,

$$\begin{aligned}\cos 2\theta &= \cos(\theta+\theta)\\ &= \cos\theta\cos\theta - \sin\theta\sin\theta\\ &= \cos^2\theta - \sin^2\theta\end{aligned}$$

We can get two more important identities for $\cos 2\theta$ by using *Theorems 7.2* and *7.3*.

Theorem 7.14: $\cos 2\theta = 2\cos^2\theta - 1$.

Proof: By *Theorem 7.13*, $\cos 2\theta = \cos^2\theta - \sin^2\theta$. Replace $\sin^2\theta$ using *Theorem 7.2*. Then $\cos 2\theta = \cos^2\theta - (1 - \cos^2\theta)$, and the identity of this theorem follows.

Theorem 7.15: $\cos 2\theta = 1 - 2\sin^2\theta$.

Proof: Again, by *Theorem 7.13*, $\cos 2\theta = \cos^2\theta - \sin^2\theta$. Replace $\cos^2\theta$ using *Theorem 7.3*. Then $\cos 2\theta = (1 - \sin^2\theta) - \sin^2\theta$, and the identity of this theorem follows.

The following theorem is an important corollary to *Theorems 7.14 and 7.15*.

Theorem 7.16: $\cos^2\theta = \dfrac{1+\cos 2\theta}{2}$ and $\sin^2\theta = \dfrac{1-\cos 2\theta}{2}$.

Example 11: Suppose $\cos\theta = \dfrac{1}{3}$ and $\sin\theta < 0$. Find $\sin 2\theta$, $\cos 2\theta$, and the quadrant where 2θ lies.

Since $\cos\theta > 0$ and $\sin\theta < 0$, the angle θ lies in the fourth quadrant by *Theorem 2.5*.

By *Theorem 7.2*, $\sin^2\theta = 1 - \left(\dfrac{1}{3}\right)^2 = 1 - \dfrac{1}{9} = \dfrac{8}{9}$.

By the Square Root Principle Theorem (see p. 166), $\sin\theta = \pm\dfrac{2\sqrt{2}}{3}$.

Since $\sin\theta < 0$, $\sin\theta = -\dfrac{2\sqrt{2}}{3}$. Now, by *Theorem 7.12*,

$$\begin{aligned}\sin 2\theta &= 2\sin\theta\cos\theta\\ &= 2\left(-\frac{2\sqrt{2}}{3}\right)\left(\frac{1}{3}\right)\\ &= -\frac{4\sqrt{2}}{9}\end{aligned}$$

By *Theorem 7.14*,

$$\begin{aligned}\cos 2\theta &= 2\cos^2\theta - 1\\ &= 2\left(\frac{1}{3}\right)^2 - 1\\ &= \frac{2}{9} - 1\\ &= -\frac{7}{9}\end{aligned}$$

Since both cos 2θ and sin 2θ are negative, 2θ lies in the third quadrant, by *Theorem 2.5*.

Example 12: Prove the identity $\sin 3\theta = 3\sin\theta - 4\sin^3\theta$.

$$
\begin{aligned}
\text{left side} &= \sin 3\theta \\
&= \sin(\theta + 2\theta) \\
&= \sin\theta\cos 2\theta + \cos\theta\sin 2\theta \\
&= \sin\theta\left(1 - 2\sin^2\theta\right) + \cos\theta\left(2\sin\theta\cos\theta\right) \\
&= \sin\theta - 2\sin^3\theta + 2\sin\theta\cos^2\theta \\
&= \sin\theta - 2\sin^3\theta + 2\sin\theta\left(1 - \sin^2\theta\right) \\
&= 3\sin\theta - 4\sin^3\theta \\
&= \text{right side}
\end{aligned}
$$

In the above proof for Example 12, we used *Theorems 7.8, 7.15, 7.12,* and *7.2.*

Example 13: Verify the identity $1 - \cos 2x = \tan x \sin 2x$.

$$
\begin{aligned}
\text{right side} &= \tan x \sin 2x \\
&= \left(\frac{\sin x}{\cos x}\right)(2\sin x\cos x) \\
&= 2\sin^2 x \\
&= 2\left(\frac{1 - \cos 2x}{2}\right) \\
&= 1 - \cos 2x \\
&= \text{left side}
\end{aligned}
$$

We used *Theorems 7.12* and *6.16* in establishing the identity in Example 13.

Theorem 7.17: $\tan 2\theta = \dfrac{2\tan\theta}{1 - \tan^2\theta}$.

Proof: Since tan 2θ = tan $(\theta + \theta)$, we can use *Theorem 7.10* to establish this identity.

$$\begin{aligned}\tan 2\theta &= \tan(\theta+\theta)\\ &= \frac{\tan\theta+\tan\theta}{1-\tan\theta\tan\theta}\\ &= \frac{2\tan\theta}{1-\tan^2\theta}\end{aligned}$$

Example 14: Find tan 2θ if $\sin\theta = \frac{2}{3}$, and $\cos\theta < 0$.

By *Theorem 7.3*,

$$\begin{aligned}\cos^2\theta &= 1-\sin^2\theta\\ &= 1-\left(\frac{2}{3}\right)^2\\ &= 1-\frac{4}{9}\\ &= \frac{5}{9}\end{aligned}$$

By the Square Root Principle Theorem (see p. 166), $\cos\theta = \pm\frac{\sqrt{5}}{3}$. However, cos θ < 0. Therefore, $\cos\theta = -\frac{\sqrt{5}}{3}$. Then $\tan\theta = \frac{\sin\theta}{\cos\theta} = -\frac{2}{\sqrt{5}}$. By *Theorem 7.17*,

$$\begin{aligned}\tan 2\theta &= \frac{2\tan\theta}{1-\tan^2\theta}\\ &= \frac{2\left(-\frac{2}{\sqrt{5}}\right)}{1-\left(-\frac{2}{\sqrt{5}}\right)^2}\\ &= -\frac{20}{\sqrt{5}}\\ &= -4\sqrt{5}\end{aligned}$$

Half-Angle Identities

The sine and cosine half-angle identities are derived by using *Theorem 7.16.*

Theorem 7.18: $\sin\left(\frac{\theta}{2}\right)=\pm\sqrt{\frac{1-\cos\theta}{2}}$. The sign of the identity depends on the quadrant in which $\frac{\theta}{2}$ is located.

Proof: By *Theorem 7.16,*

$$\sin^2 x=\frac{1-\cos 2x}{2} \qquad (1)$$

Let $2x=\theta$. Then $x=\frac{\theta}{2}$. Substitute these in (1) and we get $\sin^2\left(\frac{\theta}{2}\right)=\frac{1-\cos\theta}{2}$.

By the Square Root Principle Theorem (see p. 166), $\sin\left(\frac{\theta}{2}\right)=\pm\sqrt{\frac{1-\cos\theta}{2}}$.

Theorem 7.19: $\cos\left(\frac{\theta}{2}\right)=\pm\sqrt{\frac{1+\cos\theta}{2}}$. The sign of the identity depends on the quadrant in which $\frac{\theta}{2}$ is located.

Proof: By *Theorem 7.16,*

$$\cos^2 x=\frac{1+\cos 2x}{2} \qquad (2)$$

Let $2x=\theta$. Then $x=\frac{\theta}{2}$. Substitute these in (2) and we get $\cos^2\left(\frac{\theta}{2}\right)=\frac{1+\cos\theta}{2}$.

By the Square Root Principle Theorem (see p. 166), $\cos\left(\frac{\theta}{2}\right)=\pm\sqrt{\frac{1+\cos\theta}{2}}$.

Theorem 7.20: $\tan\left(\frac{\theta}{2}\right)=\pm\sqrt{\frac{1-\cos\theta}{1+\cos\theta}}$. The sign of the identity depends on the quadrant in which $\frac{\theta}{2}$ is located.

Theorem 7.20 follows directly by the definition of $\tan\left(\frac{\theta}{2}\right)$ and *Theorems 7.18* and *7.19*.

Theorem 7.21: $\tan\frac{\theta}{2}=\frac{\sin\theta}{1+\cos\theta}$.

Proof: By *Theorem 7.12*, $\sin 2x = 2\sin x\cos x$. Therefore,

$$\sin x=\frac{\sin 2x}{2\cos x} \qquad (1)$$

By *Theorem 7.16*, $\cos^2 x=\frac{1+\cos 2x}{2}$. Therefore,

$$\cos x=\frac{1+\cos 2x}{2\cos x} \qquad (2)$$

By dividing (1) by (2),

$$\tan x=\frac{\sin 2x}{1+\cos 2x} \qquad (3)$$

Now, let $2x = \theta$. Then $x=\frac{\theta}{2}$. Substitute these into (3):

$$\tan\frac{\theta}{2}=\frac{\sin\theta}{1+\cos\theta}$$

Theorem 7.22: $\tan\frac{\theta}{2}=\frac{1-\cos\theta}{\sin\theta}$.

Proof: By *Theorem 7.16*, $\sin^2 x=\frac{1-\cos 2x}{2}$. Therefore,

$$\sin x=\frac{1-\cos 2x}{2\sin x} \qquad (1)$$

By *Theorem 7.12*, $\sin 2x = 2\sin x\cos x$. Therefore,

$$\cos x=\frac{\sin 2x}{2\sin x} \qquad (2)$$

By dividing (1) by (2),

$$\tan x = \frac{1-\cos 2x}{\sin 2x} \qquad (3)$$

Now, let $2x = \theta$. Then $x = \frac{\theta}{2}$. Substitute these into (3):

$$\tan\frac{\theta}{2} = \frac{1-\cos\theta}{\sin\theta}$$

Example 15: Find the exact value of sin (22.5°).

Since $22.5° = \frac{45°}{2}$, and 22.5° is in the first quadrant, we will use *Theorem 7.18.*

$$\begin{aligned}\sin(22.5°) &= \sqrt{\frac{1-\cos 45°}{2}}\\ &= \sqrt{\frac{1-\frac{\sqrt{2}}{2}}{2}}\\ &= \frac{\sqrt{2-\sqrt{2}}}{2}\end{aligned}$$

Example 16: Find the exact value of tan (7.5°).

Since $7.5° = \frac{15°}{2}$, we will first find sin 15° and cos 15°, then use *Theorem 7.21.*

By *Theorem 7.18,*

$$\begin{aligned}\sin 15° &= \sqrt{\frac{1-\cos 30°}{2}}\\ &= \sqrt{\frac{1-\frac{\sqrt{3}}{2}}{2}}\\ &= \frac{\sqrt{2-\sqrt{3}}}{2}\end{aligned}$$

By *Theorem 7.19*,

$$\cos 15^\circ = \sqrt{\frac{1+\cos 30^\circ}{2}}$$

$$= \sqrt{\frac{1+\frac{\sqrt{3}}{2}}{2}}$$

$$= \frac{\sqrt{2+\sqrt{3}}}{2}$$

Then by *Theorem 7.21*,

$$\tan(7.5^\circ) = \frac{\frac{\sqrt{2-\sqrt{3}}}{2}}{1+\frac{\sqrt{2+\sqrt{3}}}{2}}$$

$$= \frac{\sqrt{2-\sqrt{3}}}{2+\sqrt{2+\sqrt{3}}}$$

Product-to-Sum and Sum-to-Product Identities

The process of converting sums into products or products into sums can make the difference between an easy solution to a problem and no solution at all. Two sets of identities that can be derived from the sum and difference identities will be helpful in this conversion.

The following identities are known as the *product-to-sum identities.*

Theorem 7.23 (Product-to-Sum Identities Theorem):

$$(1)\ \cos\alpha\cos\beta = \frac{1}{2}\left[\cos(\alpha+\beta)+\cos(\alpha-\beta)\right]$$

$$(2)\ \sin\alpha\sin\beta = -\frac{1}{2}\left[\cos(\alpha+\beta)-\cos(\alpha-\beta)\right]$$

$$(3)\ \sin\alpha\cos\beta = \frac{1}{2}\left[\sin(\alpha+\beta)+\sin(\alpha-\beta)\right]$$

$$(4)\ \cos\alpha\sin\beta = \frac{1}{2}\left[\sin(\alpha+\beta)-\sin(\alpha-\beta)\right]$$

Proof: From *Theorems 7.6–7.9*, the following are sum and difference identities:

$$\cos(\alpha-\beta)=\cos\alpha\cos\beta+\sin\alpha\sin\beta \quad \text{(A)}$$
$$\cos(\alpha+\beta)=\cos\alpha\cos\beta-\sin\alpha\sin\beta \quad \text{(B)}$$
$$\sin(\alpha+\beta)=\sin\alpha\cos\beta+\cos\alpha\sin\beta \quad \text{(C)}$$
$$\sin(\alpha-\beta)=\sin\alpha\cos\beta-\cos\alpha\sin\beta \quad \text{(D)}$$

(B) + (A) gives $\cos(\alpha+\beta)+\cos(\alpha-\beta)=2\cos\alpha\cos\beta$. Divide both sides by 2 and we get (1).

(B) – (A) gives $\cos(\alpha+\beta)-\cos(\alpha-\beta)=-2\sin\alpha\sin\beta$. Divide both sides by 2 and we get (2).

(C) + (D) gives $\sin(\alpha+\beta)+\sin(\alpha-\beta)=2\sin\alpha\cos\beta$. Divide both sides by 2 and we get (3).

(C) – (D) gives $\sin(\alpha+\beta)-\sin(\alpha-\beta)=2\cos\alpha\sin\beta$. Divide both sides by 2 and we get (4).

Example 17: Write cos $3x$ cos $2x$ as a sum.

$$\cos\alpha\cos\beta=\frac{1}{2}\left[\cos(\alpha+\beta)+\cos(\alpha-\beta)\right]$$
$$\cos 3x\cos 2x=\frac{1}{2}\left[\cos(3x+2x)+\cos(3x-2x)\right]$$
$$\cos 3x\cos 2x=\frac{1}{2}(\cos 5x+\cos x)$$
$$\cos 3x\cos 2x=\frac{\cos 5x}{2}+\frac{\cos x}{2}$$

Alternate forms of the product-to-sum identities are the *sum-to-product identities.*

Theorem 7.24 (Sum-to-Product Identities Theorem):

$$(1)\cos x+\cos y=2\cos\frac{x+y}{2}\cos\frac{x-y}{2}$$

$$(2)\cos x-\cos y=-2\sin\frac{x+y}{2}\sin\frac{x-y}{2}$$

$$(3)\sin x+\sin y=2\sin\frac{x+y}{2}\cos\frac{x-y}{2}$$

$$(4)\sin x-\sin y=2\cos\frac{x+y}{2}\sin\frac{x-y}{2}$$

Proof: From *Theorems 7.6–7.9*, the following are sum and difference identities:

$$\cos(\alpha-\beta)=\cos\alpha\cos\beta+\sin\alpha\sin\beta \quad (A)$$

$$\cos(\alpha+\beta)=\cos\alpha\cos\beta-\sin\alpha\sin\beta \quad (B)$$

$$\sin(\alpha+\beta)=\sin\alpha\cos\beta+\cos\alpha\sin\beta \quad (C)$$

$$\sin(\alpha-\beta)=\sin\alpha\cos\beta-\cos\alpha\sin\beta \quad (D)$$

Let $x = \alpha + \beta$ and $y = \alpha - \beta$. By solving this system of two equations for α and β, we get $\alpha = \frac{x+y}{2}$ and $\beta = \frac{x-y}{2}$. Substituting these values of x, y, α, and β into (A), (B), (C), and (D), we get:

$$\cos y=\cos\left(\frac{x+y}{2}\right)\cos\left(\frac{x-y}{2}\right)+\sin\left(\frac{x+y}{2}\right)\sin\left(\frac{x-y}{2}\right) \quad (A)$$

$$\cos x=\cos\left(\frac{x+y}{2}\right)\cos\left(\frac{x-y}{2}\right)-\sin\left(\frac{x+y}{2}\right)\sin\left(\frac{x-y}{2}\right) \quad (B)$$

$$\sin x=\sin\left(\frac{x+y}{2}\right)\cos\left(\frac{x-y}{2}\right)+\cos\left(\frac{x+y}{2}\right)\sin\left(\frac{x-y}{2}\right) \quad (C)$$

$$\sin y=\sin\left(\frac{x+y}{2}\right)\cos\left(\frac{x-y}{2}\right)-\cos\left(\frac{x+y}{2}\right)\sin\left(\frac{x-y}{2}\right) \quad (D)$$

(B) + (A) gives (1), (B) – (A) gives (2), (C) + (D) gives (3), and (C) – (D) gives (4).

Example 18: Write the difference cos 8α – cos 2α as a product.

$$\cos\alpha - \cos\beta = -2\sin\frac{\alpha+\beta}{2}\sin\frac{\alpha-\beta}{2}$$
$$\cos 8\alpha - \cos 2\alpha = -2\sin 5\alpha \sin 3\alpha$$

Example 19: Find the exact value of sin 75° + sin 15°.

$$\sin x + \sin y = 2\sin\frac{x+y}{2}\cos\frac{x-y}{2}$$
$$\sin 75° + \sin 15° = 2\sin 45° \cos 30°$$
$$\sin 75° + \sin 15° = 2\left(\frac{1}{\sqrt{2}}\right)\left(\frac{\sqrt{3}}{2}\right)$$
$$\sin 75° + \sin 15° = \frac{\sqrt{3}}{\sqrt{2}}$$
$$\sin 75° + \sin 15° = \frac{\sqrt{6}}{2}$$

Trigonometric Equations

We have looked at the solutions of basic trigonometric equations in Chapter 5. The following are the major findings of Chapter 5.

The general solution of sin $\theta = a$ is $n\pi + (-1)^n \sin^{-1} a$, where n is an integer. (*Theorem 5.1*)

The general solution of cos $\theta = a$ is $2n\pi \pm \cos^{-1} a$, where n is an integer. (*Theorem 5.2*)

The general solution of the equation tan $\theta = a$ is $n\pi + \tan^{-1}(a)$, where n is an integer. (*Theorem 5.3*)

We can find solutions to some other trigonometric equations based on the knowledge that we have acquired so far. Take a look at Examples 20–23.

Example 20: Find the general solution of $\cos^2 x - 2\cos x - 3 = 0$, and then find the solutions in the interval $[0, 2\pi]$.

The given equation is a quadratic equation in $\cos x$. Therefore, we can use the techniques of solving quadratic equations that we learned in algebra. The following are some of the theorems from algebra.

Theorem 7.25 (Zero-Product Property Theorem): Suppose a and b are two real numbers. If $ab = 0$, then either $a = 0$ or $b = 0$.

Theorem 7.26 (Squares Are Non-Negative Theorem): If x is a real number, then $x^2 \geq 0$.

Theorem 7.27 (Square Root Principle Theorem): Suppose $x^2 = a$ for some nonnegative real number a. Then $x = \pm\sqrt{a}$.

Theorem 7.28 (Quadratic Formula Theorem): The solutions of the equation $ax^2 + bx + c = 0$, where a, b, and c are real constants, are

$$x = \frac{-b \pm \sqrt{b^2 - 4ac}}{2a}, \text{ if } b^2 - 4ac \geq 0.$$

With this background in place, we should be able to solve the equation in Example 20.

The left side of this equation is factorable. Therefore, $\cos^2 x - 2\cos x - 3 = 0$ implies that $(\cos x - 3)(\cos x + 1) = 0$.

By the Zero-Product Property Theorem, $\cos x - 3 = 0$ or $\cos x + 1 = 0$. Therefore, $\cos x = 3$ or $\cos x = -1$.

The first equation makes no sense since $\cos x$ cannot be more than 1 (*Theorem 3.2*). The second equation is a basic cosine equation and, therefore, we know the general solution by *Theorem 5.2*: $x = 2n\pi \pm \cos^{-1}(-1)$, where n is an integer. Since $\cos^{-1}(-1) = \pi$, we get the general solution of the given equation: $x = 2n\pi \pm \pi$.

When $n = 0$, we get two solutions: $\pm\pi$. Only the solution π is in the interval $[0, 2\pi]$.

When $n = 1$, we get two solutions: $2\pi \pm \pi$. Again, only the solution π is in the given interval.

When $n = 2$, we get two solutions: $4\pi \pm \pi$. Both are outside of the given interval. The solutions are larger, for larger values of n.

When $n = -1$, we get two solutions: $-2\pi \pm \pi$. Both solutions are outside the given interval, and smaller values of n produce smaller solutions.

Therefore, the only solution of the given equation that lies in the interval $[0, 2\pi]$ is π.

Example 21: Find the general solution of $2\sin^2 x - \sin x - 1 = 0$, and then find the solutions in the interval $[0, 2\pi]$.

The equation $2\sin^2 x - \sin x - 1 = 0$ implies $(2\sin x + 1)(\sin x - 1) = 0$.

Therefore, $\sin x = -\frac{1}{2}$ or $\sin x = 1$. By *Theorem 5.1*,

$x = n\pi + (-1)^n \sin^{-1}\left(-\frac{1}{2}\right)$ or $x = n\pi + (-1)^n \sin^{-1}(1)$, where n is an integer.

We have found our general solution. It has two parts:

$$x = n\pi + (-1)^n\left(-\frac{\pi}{6}\right) \text{ or } x = n\pi + (-1)^n \frac{\pi}{2}$$

When $n = 0$, we get two solutions: $-\frac{\pi}{6}$ and $\frac{\pi}{2}$. The solution $\frac{\pi}{2}$ is in the given interval.

When $n = 1$, we get two solutions: $\pi + \frac{\pi}{6}$ and $\pi - \frac{\pi}{2}$. Both solutions are in the given interval.

When $n = 2$, we get two solutions: $2\pi - \frac{\pi}{6}$ and $2\pi + \frac{\pi}{2}$. The solution $2\pi - \frac{\pi}{6}$ is in the given interval.

Therefore, the solutions of the equations that lie in the interval $[0, 2\pi]$ are $\frac{\pi}{2}$, $\frac{7\pi}{6}$, and $\frac{11\pi}{6}$.

Example 22: Find the general solution of the equation $\tan^2 x - 1 = 0$, and then find the solutions in the interval $[0, 2\pi]$.

The equation $\tan^2 x - 1 = 0$ implies that $\tan^2 x = 1$. Then by the Square Root Principle Theorem, $\tan x = \pm 1$.

By *Theorem 5.3*, $x = n\pi + \tan^{-1}(1)$ or $x = n\pi + \tan^{-1}(-1)$, where n is an integer.

We have found our general solution. It has two parts:

$$x = n\pi + \frac{\pi}{4} \text{ or } x = n\pi - \frac{\pi}{4}$$

When $n = 0$, we get two solutions: $\frac{\pi}{4}$ and $-\frac{\pi}{4}$. The solution $\frac{\pi}{4}$ is in the given interval.

When $n = 1$, we get two solutions: $\pi + \frac{\pi}{4}$ and $\pi - \frac{\pi}{4}$. Both solutions are in the given interval.

When $n = 2$, we get two solutions: $2\pi + \frac{\pi}{4}$ and $2\pi - \frac{\pi}{4}$. The solution $2\pi - \frac{\pi}{4}$ is in the given interval.

Therefore, the solutions of the equations that lie in the interval $[0, 2\pi]$ are $\frac{\pi}{4}$, $\frac{3\pi}{4}$, $\frac{5\pi}{4}$, and $\frac{7\pi}{4}$.

Example 23: Find the general solution of the equation $\sin x \cos x = \frac{1}{4}$, and then find the solutions in the interval $[0, 2\pi]$.

We can use the double-angle identity in *Theorem 7.12* to solve this equation. $\sin x \cos x = \frac{1}{4}$ implies that $2\sin x \cos x = \frac{1}{2}$. That in turn implies that $\sin 2x = \frac{1}{2}$. This is a basic sine equation in $2x$. Therefore, by *Theorem 5.1*, $2x = n\pi + (-1)^n \sin^{-1}\left(\frac{1}{2}\right)$, where n is an integer. That is, $2x = n\pi + (-1)^n \frac{\pi}{6}$. Then the general solution to the equation is $x = \frac{n\pi}{2} + (-1)^n \frac{\pi}{12}$.

When $n = 0$, we get $\frac{\pi}{12}$, which is in the given interval.

When $n = 1$, we get $\frac{\pi}{2} - \frac{\pi}{12}$, which is in the given interval.

When $n = 2$, we get $\pi + \frac{\pi}{12}$, which is in the given interval.

When $n = 3$, we get $\frac{3\pi}{2} + \frac{\pi}{12}$, which is in the given interval.

When $n = 4$, we get $2\pi - \frac{\pi}{12}$, which is in the given interval.

Therefore, the solutions of the equations that lie in the interval $[0, 2\pi]$ are $\frac{\pi}{12}$, $\frac{5\pi}{12}$, $\frac{13\pi}{12}$, $\frac{19\pi}{12}$, and $\frac{23\pi}{12}$.

Chapter Check-Out

Questions

1. Use the fundamental trigonometric identities to find the remaining five trig functions if $\cos \alpha = \frac{2}{3}$ and $\tan \theta < 0$.
2. Establish this identity: $\sec \theta - \cos \theta = \tan \theta \sin \theta$.
3. True or False: The sum and difference formulas can be used to find the exact value of the cosine of 75°.
4. Establish this identity: $\frac{\cos\theta + \sin\theta}{\sin\theta} = 1 + \cot\theta$.
5. Find the general solution of the equation $\cos 2x = 0$.

Answers

1. $\sin\theta = -\frac{\sqrt{5}}{3}$, $\tan\theta = -\frac{\sqrt{5}}{2}$, $\cot\theta = -\frac{2\sqrt{5}}{5}$, $\sec\theta = \frac{3}{2}$, $\csc\theta = -\frac{3\sqrt{5}}{5}$

2.
$$\begin{aligned}
\text{left side} &= \sec\theta - \cos\theta \\
&= \frac{1}{\cos\theta} - \cos\theta \\
&= \frac{1 - \cos^2\theta}{\cos\theta} \\
&= \frac{\sin^2\theta}{\cos\theta} \\
&= \frac{\sin\theta \sin\theta}{\cos\theta} \\
&= \tan\theta \sin\theta \\
&= \text{right side}
\end{aligned}$$

3. True
4. $$\begin{aligned}\text{left side} &= \frac{\cos\theta + \sin\theta}{\sin\theta} \\ &= \frac{\cos\theta}{\sin\theta} + \frac{\sin\theta}{\sin\theta} \\ &= \cot\theta + 1 \\ &= \text{right side}\end{aligned}$$
5. $n\pi \pm \frac{\pi}{4}$, where n is an integer

Chapter 8
VECTORS

Chapter Check-In

- ❑ Understanding vectors and their components
- ❑ Understanding vector operations
- ❑ Using vectors in a rectangular coordinate system to solve problems
- ❑ Understanding the dot product of two vectors

Common Core Standard

Please refer to the standards for mathematical practice, found here: www.corestandards.org/Math/Practice/.

In the physical world, some quantities, such as mass, length, and area, can be represented by a number. These numbers are known as **scalars.** Other quantities, such as velocity and force, also involve a direction. Such a quantity can be represented by a mathematical concept called a **vector** with only two properties: the magnitude and the direction. One common use of vectors involves finding the actual speed and direction of an aircraft when given its airspeed and direction and the speed and direction of a tailwind. Another common use of vectors involves finding the resulting force on an object being acted on by several separate forces.

Trigonometry Theorems

The trigonometry theorems in this book are numbered for organizational purposes. For example, *Theorem 2.2* is the second theorem given in Chapter 2. Study these theorems by their content, not by their number, as the theorem numbering has no significance outside of this book.

The discussion in this chapter is limited to vectors in a two-dimensional coordinate plane, although the concepts can be extended to higher dimensions.

To define a vector, we can use a line segment with endpoints identified as the **initial point** and the **terminal point.** The length of the line segment represents the **magnitude of the vector,** and looking at this segment as an arrow starting from the initial point and ending at the terminal point gives us the **direction of the vector.**

However, we want to broaden the meaning of the direction based on physical quantities that we are going to use vectors to represent. Imagine two cars traveling in parallel lanes on a straight highway. We say they travel in the "same" direction. (The length of a vector, in this context, represents the speed of a car.) Then we say the vectors that represent their velocities have the "same" direction.

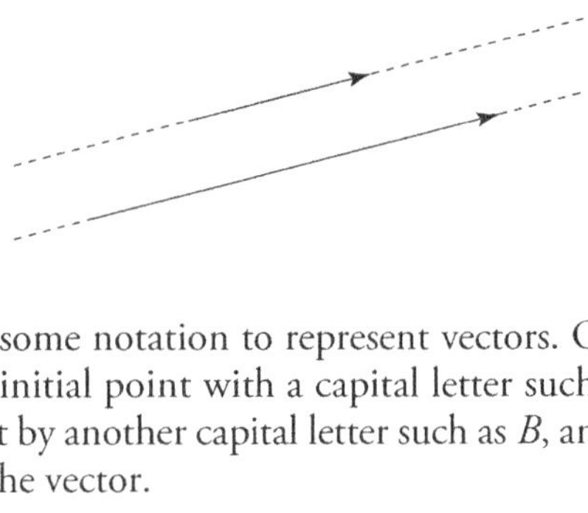

Clearly, we need some notation to represent vectors. One way to do this is to identify the initial point with a capital letter such as A and identify the terminal point by another capital letter such as B, and use the notation $\overrightarrow{AB}$ to represent the vector.

Unfortunately, this is also the notation used to represent a ray with initial point A and passing through point B. As long as it is understood that we are using this notation to represent a vector now, there shouldn't be any confusion.

However, we will also use another, perhaps better, notation to represent vectors: a lowercase bold letter. The letters **u**, **v**, and **w** are frequently used to represent vectors.

If we interchange the initial point and the terminal point of a vector, then we get a new vector. We say the second vector has the direction opposite to the first vector.

In the following figure, the vectors **u** and **v** have opposite directions.

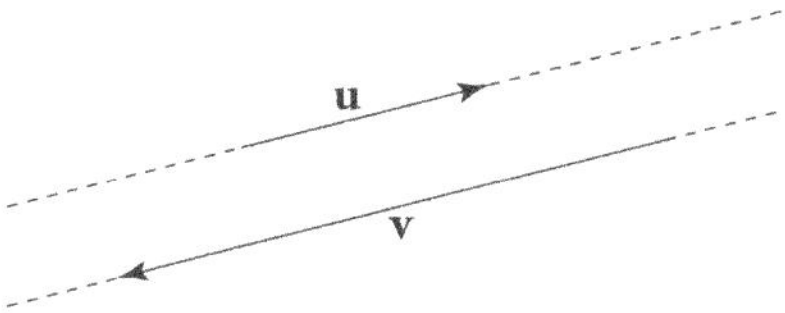

We say vector **u** and **v** are **equal vectors** if they have the same magnitude and the same direction. Since a vector has only two properties, magnitude and direction, we say **u** and **v** represent the *same* vector in this case. That is, we can move a vector as long as we do not change its magnitude and the direction. In that sense, the following are the same vector.

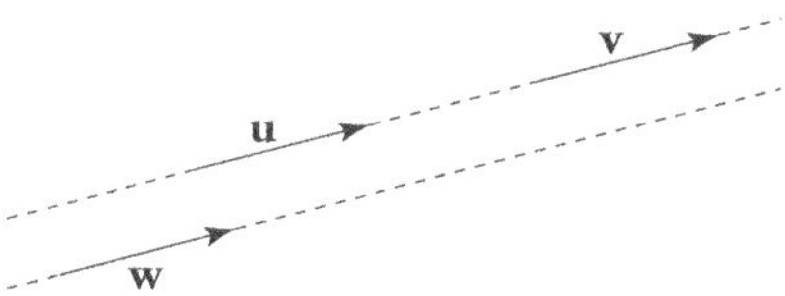

Vector Operations

A vector that has no magnitude and points in any direction is called a **zero vector.** We denote this vector as **0**.

We will define **vector addition** as follows. Let **u** and **v** be two vectors. Move the vector **v** so that its initial point is the same as the terminal point of **u**. The vector with the initial point the same as the initial point of **u** and the terminal point the same as **v** is called **u** + **v**.

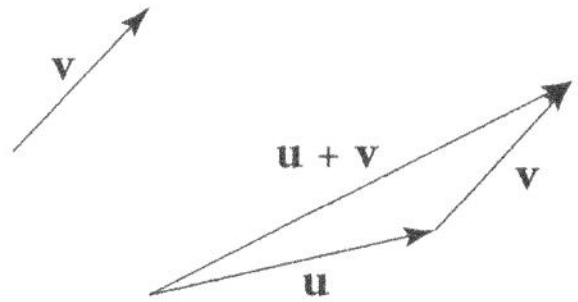

We will define **v** + **u** the same way. Move the vector **u** so that its initial point is the same as the terminal point of **v**. The vector with the initial point the same as the initial point of **v** and the terminal point the same as **u** is called **v** + **u**.

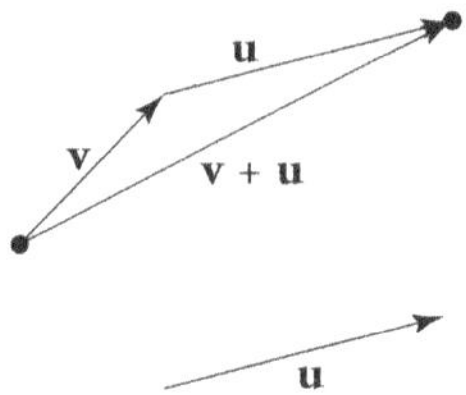

Does the commutative property hold for vector addition? The answer is yes.

Theorem 8.1: If **u** and **v** are two vectors, then **u** + **v** = **v** + **u**.

Proof:

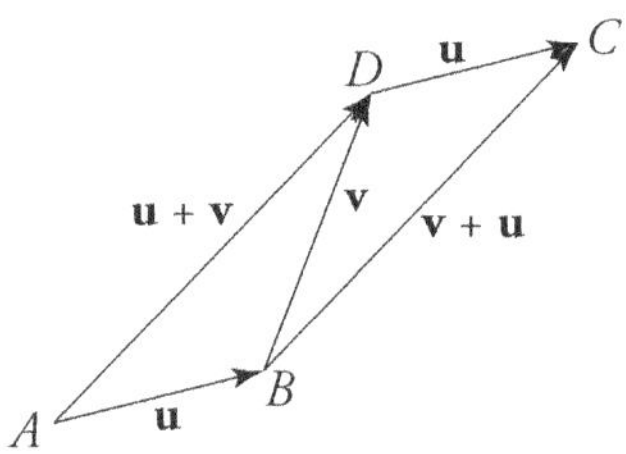

Consider the quadrilateral *ABCD*, where the vector $\overrightarrow{AB} = \mathbf{u}$, $\overrightarrow{BD} = \mathbf{v}$, and $\overrightarrow{DC} = \mathbf{u}$. Then by the definition of vector addition, $\overrightarrow{AD} = \mathbf{u} + \mathbf{v}$ and $\overrightarrow{BC} = \mathbf{v} + \mathbf{u}$.

$\overline{AB} = \overline{DC}$ because these lengths are the magnitudes of the same vector. Line $\overleftrightarrow{AB}$ is parallel to line $\overleftrightarrow{DC}$ because vector **u** lies on those lines. Therefore, *ABCD* is a parallelogram. (If two opposite sides of a quadrilateral are equal and parallel, then the quadrilateral is a parallelogram. See *CliffsNotes Geometry Common Core Quick Review*.) Therefore, $\overline{AD} = \overline{BC}$, and the lines $\overleftrightarrow{AD}$ and $\overleftrightarrow{BC}$ are parallel. (In a parallelogram, any two opposite sides are equal and parallel. See *CliffsNotes Geometry Common Core Quick Review*.) Therefore, **u** + **v** = **v** + **u**.

We will denote the magnitude of a vector **u** by the notation $\|\mathbf{u}\|$. The magnitude of a vector is a nonnegative number. The magnitude is equal to zero for the zero vector.

We define another operation for vectors called **scalar multiplication** as follows. Remember that a scalar is a number. Let k be a scalar and **u** be a vector. Since k is a number, the trichotomy law holds. That is, k is positive, negative, or zero. If $k > 0$, then we define $k\mathbf{u}$ as the vector with the same direction as **u** and the magnitude $k\|\mathbf{u}\|$. If $k < 0$, then we define the vector $k\mathbf{u}$ as the vector with the direction opposite to $\mathbf{u}$ and the magnitude $|k|\,\|\mathbf{u}\|$. If $k = 0$, then $k\mathbf{u} = 0$.

Example 1: Consider the vector given in the following figure with magnitude 4. Sketch the vectors $\frac{1}{2}\mathbf{u}$, $2\mathbf{u}$, and $-1\mathbf{u}$.

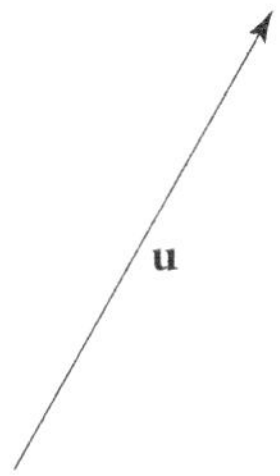

The following figure shows vectors $\mathbf{u}$, $\frac{1}{2}\mathbf{u}$, $2\mathbf{u}$, and $-1\mathbf{u}$ drawn according to the definition of the scalar multiplication of a vector.

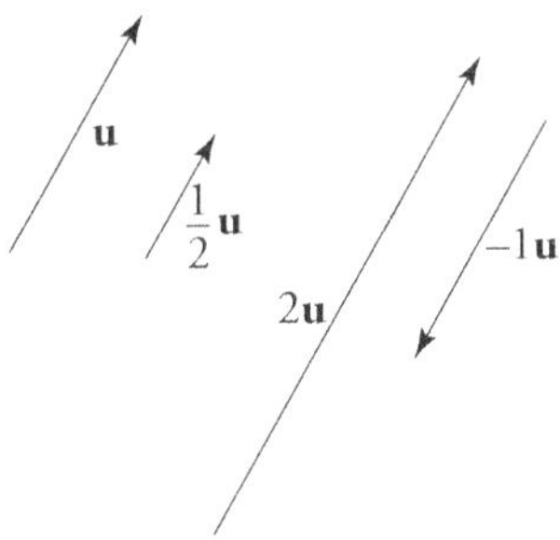

The following theorem follows from the definition of the scalar multiplication of a vector.

Theorem 8.2: If m and n are scalars and $\mathbf{u}$ is a vector, then $m(n\mathbf{u}) = n(m\mathbf{u}) = (mn)\mathbf{u}$.

Proof: If one of m or n is 0, then by the definition of the scalar multiplication of a vector, $m(n\mathbf{u}) = n(m\mathbf{u}) = (mn)\mathbf{u} = \mathbf{0}$.

If both m and n have the same sign, then then by the definition of the scalar multiplication, all three vectors $m(n\mathbf{u})$, $n(m\mathbf{u})$, $(mn)\mathbf{u}$ have the same direction as $\mathbf{u}$ and has magnitude $mn\|\mathbf{u}\|$. If m and n have opposite signs, then by the definition of the scalar multiplication, all three vectors—$m(n\mathbf{u})$, $n(m\mathbf{u})$, and $(mn)\mathbf{u}$—have the same direction as $-\mathbf{u}$ and have magnitude $|mn|\,\|\mathbf{u}\|$.

We define $-\mathbf{u}$ as the vector $-1\mathbf{u}$. We also define $\mathbf{u} - \mathbf{v}$ as the vector $\mathbf{u} + (-\mathbf{v})$.

A vector with magnitude 1 is called a **unit vector.**

Theorem 8.3: Let $\mathbf{u}$ be a non-zero vector. Then $\frac{1}{\|\mathbf{u}\|}\mathbf{u}$ is a unit vector.

Proof: Since $\mathbf{u}$ is a non-zero vector, $\|\mathbf{u}\| \neq 0$. Therefore, $\frac{1}{\|\mathbf{u}\|}$ is a positive scalar. By the definition of the scalar multiplication of a vector, the magnitude of $\frac{1}{\|\mathbf{u}\|}\mathbf{u}$ is $\frac{1}{\|\mathbf{u}\|}\|\mathbf{u}\|$. Since $\frac{1}{\|\mathbf{u}\|}\|\mathbf{u}\| = 1$, $\frac{1}{\|\mathbf{u}\|}\mathbf{u}$ is a unit vector.

Theorem 8.4: Suppose $\mathbf{u}$ is a non-zero vector. Then $\|\mathbf{u}\|\left(\frac{1}{\|\mathbf{u}\|}\mathbf{u}\right) = \mathbf{u}$.

Proof: $\|\mathbf{u}\|\left(\frac{1}{\|\mathbf{u}\|}\mathbf{u}\right) = \left(\|\mathbf{u}\|\frac{1}{\|\mathbf{u}\|}\right)\mathbf{u} = \mathbf{u}$. The first equality follows from *Theorem 8.2.*

Since a vector has only two properties, the magnitude and the direction, and the magnitude is $\|\mathbf{u}\|$ as a consequence of *Theorem 8.4*, we can consider the unit vector $\frac{1}{\|\mathbf{u}\|}\mathbf{u}$ as the "direction" of a non-zero vector $\mathbf{u}$.

Let $\overrightarrow{AB} = \mathbf{u}$ be a non-zero vector with initial point A and the terminal point B, and let $\overrightarrow{AC} = \mathbf{v}$ be a non-zero vector with initial point A and the terminal point C. (Since we can move vectors, this arrangement is always possible for two non-zero vectors.)

The angle θ formed by the two rays $\overrightarrow{AB}$ and $\overrightarrow{AC}$, where $0 \leq \theta \leq \pi$, is called the angle between $\mathbf{u}$ and $\mathbf{v}$. When $\mathbf{u}$ and $\mathbf{v}$ have the same direction, $\theta = 0$, and when $\mathbf{u}$ and $\mathbf{v}$ have opposite directions, $\theta = \pi$.

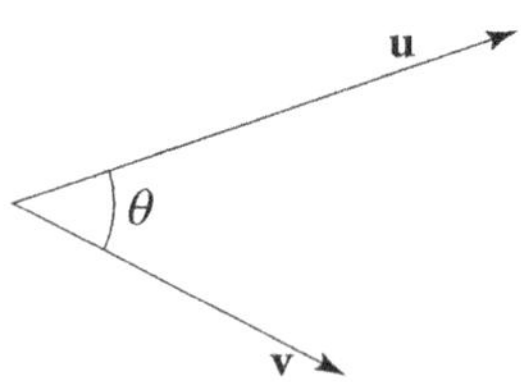

Two non-zero vectors **u** and **v** are **orthogonal** if the angle between them is $\frac{\pi}{2}$.

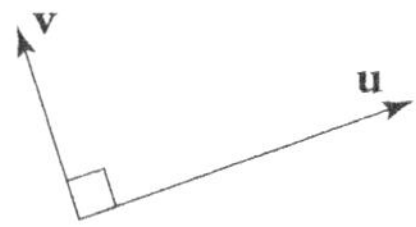

Theorem 8.5: Let **u** and **v** be two non-zero vectors. Then the vector **u** can be written as a sum of two vectors orthogonal to each other so that one of the two vectors has the same direction as **v** or –**v**.

The vector with the same direction as **v** or –**v** is called the vector component of **u** in the direction of **v** and the other vector is called the vector component of **u** orthogonal to **v**.

Proof: If the angle between **u** and **v** is either 0 or π, then the vector component of **u** in the direction of **v** is **u**, and the vector component of **u** orthogonal to **v** is **0**.

If the angle θ between **u** and **v** is $\frac{\pi}{2}$, then the vector component of **u** in the direction of **v** is **0**, and the vector component of **u** orthogonal to **v** is **u**.

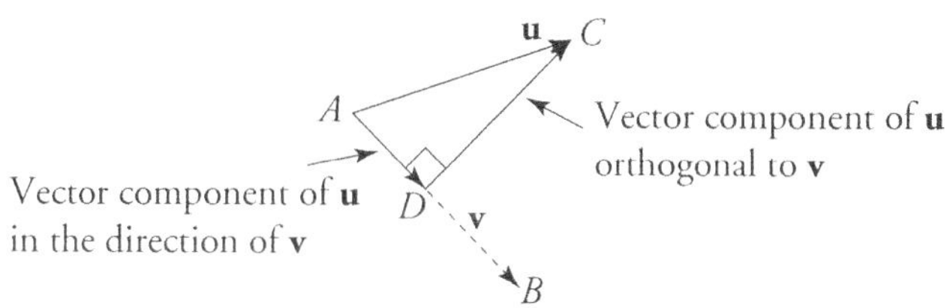

If $0 < \theta < \pi$, then let $\overrightarrow{AB} = \mathbf{v}$ and $\overrightarrow{AC} = \mathbf{u}$. Draw a perpendicular from C to line $\overleftrightarrow{AB}$, and suppose D is the foot of the perpendicular. Then the vector $\overrightarrow{AD}$ is the vector component of **u** in the direction of **v**, and the vector $\overrightarrow{DC}$ is the vector component of **u** orthogonal to **v**.

Example 2: A plane is traveling due west with an airspeed of 400 miles per hour. There is a tailwind blowing in a southwest direction at 50 miles per hour. Draw a diagram that represents the plane's ground speed and direction.

Let the unit vector in the direction of west be **u** and let the unit vector in the direction of southwest be **v**.

Then the actual velocity vector of the plane is 400**u** and the velocity vector of the tailwind is 50**v**. By the definition of vector addition, a person observing the plane sees the plane traveling in the direction of the (resulting) vector **w** = 400**u** + 50**v**. The vector **w** is the resultant **velocity** of the plane. See the figure that follows.

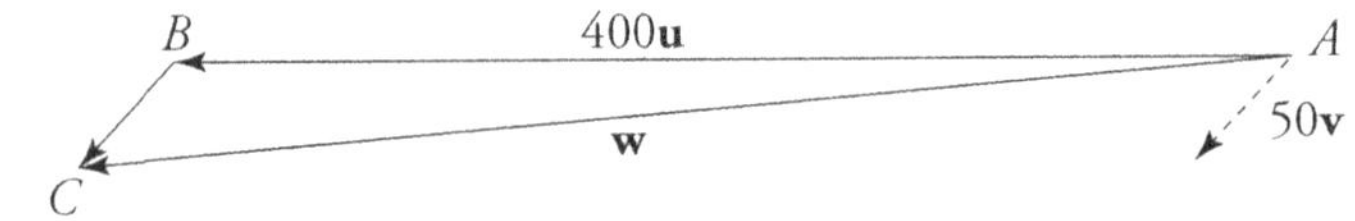

The **bearing** of a vector **w** is the angle measured clockwise from due north to **w**. In Example 2, the bearing of the vector 400**u** is 270°, and the bearing of 50**v** is 225°. Using this information, we can calculate the magnitude of **w** and the bearing θ of **w**.

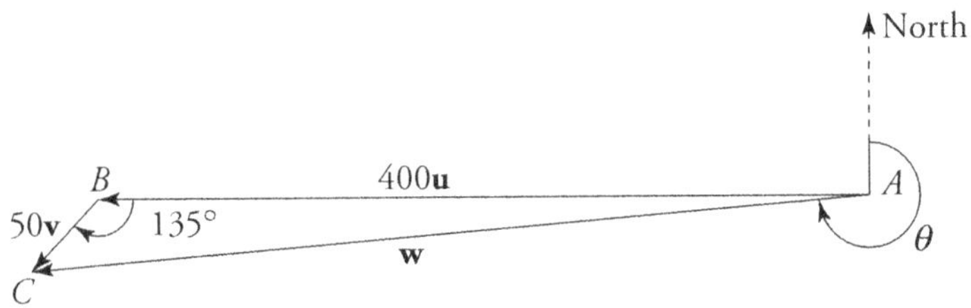

Using the Law of Cosines Theorem on ΔABC in the above figure,

$$\begin{aligned}\|\mathbf{w}\|^2 &= \|400\mathbf{u}\|^2 + \|50\mathbf{v}\|^2 - 2(\|400\mathbf{u}\|)(\|50\mathbf{v}\|)\cos 135^\circ \\ &= 400^2 + 50^2 - 2(400)(50)\left(-\frac{\sqrt{2}}{2}\right) \\ &= 50^2\left(65 + 8\sqrt{2}\right)\end{aligned}$$

Since $\|\mathbf{w}\|$ is a positive number,

$$\begin{aligned}\|\mathbf{w}\| &= 50\sqrt{65+8\sqrt{2}} \\ &\approx 436.8\end{aligned}$$

Therefore, the resultant speed of the plane is approximately 436.8 miles per hour.

Let the measure of $\angle BAC$ be α. Then, by the Law of Sines Theorem, $\frac{\|\mathbf{w}\|}{\sin 135^\circ} = \frac{50}{\sin\alpha}$ and $\sin\alpha = \frac{50\sin 135^\circ}{\|\mathbf{w}\|}$.

This equation has two solutions in the interval (0°, 180°). However, $\angle BAC$ has to be acute by *Theorem 6.10*, since $\angle ABC$ has the measure 135°. Therefore,

$$\alpha = \sin^{-1}\left(\frac{50\sin 135^\circ}{\|\mathbf{w}\|}\right)$$
$$\approx 4.64^\circ$$

The bearing, θ, is therefore 270° – 4.64°, or approximately 265.4°.

Example 3: A plane flies at 300 miles per hour in some direction. There is a wind blowing southeast at 86 miles per hour with a bearing of 320°. At what bearing must the plane head in order to have a true bearing (relative to the ground) of 14°? What will be the plane's ground speed?

Let **u** be the unit vector in the direction of the velocity of the plane, let **v** be the unit vector in the direction of the velocity of the wind, and let **w** be the resultant velocity of the plane. Then $\mathbf{w} = 300\mathbf{u} + 86\mathbf{v}$.

Let α be the measure of the angle between the vectors **w** and **u**. Consider ΔABC shown in the following figure; we will use the standard convention to identify the angle measures of ΔABC. That is, $\angle ACB$ has measure γ and $\angle ABC$ has measure β. Then $\gamma = 54°$.

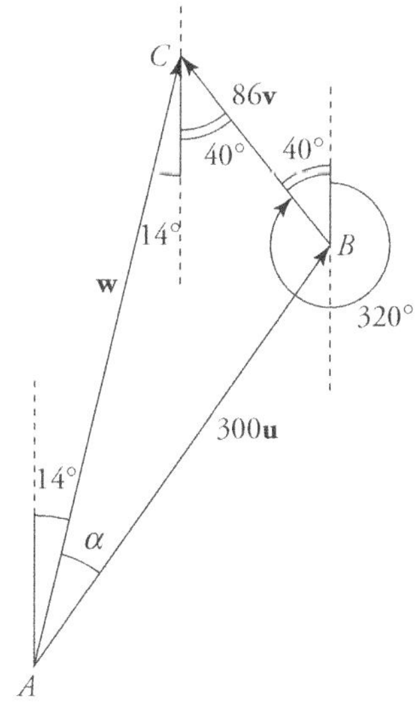

By the Law of Sines Theorem,

$$\frac{\|86\mathbf{v}\|}{\sin\alpha} = \frac{\|300\mathbf{u}\|}{\sin 54^\circ}$$
$$\frac{86}{\sin\alpha} = \frac{300}{\sin 54^\circ}$$
$$\sin\alpha = \frac{43\sin 54^\circ}{150}$$

Since α is an angle of a triangle, α is in the interval (0°, 180°). The above basic sine equation has two solutions in (0°, 180°): $\alpha = \sin^{-1}\left(\frac{43\sin 54^\circ}{150}\right)$ or $\alpha = 180^\circ - \sin^{-1}\left(\frac{43\sin 54^\circ}{150}\right)$.

These solutions are approximately $\alpha \approx 13.41^\circ$ or $\alpha \approx 166.59^\circ$. However, 300 > 86. Therefore, by *Theorem 6.4*, α has to be less than 54°. Therefore, $\alpha \approx 13.41^\circ$.

Now by *Theorem 6.10*, $\beta \approx 112.59^\circ$. By using the Law of Sines Theorem again,

$$\frac{\|\mathbf{w}\|}{\sin\beta} = \frac{\|300\mathbf{u}\|}{\sin 54^\circ}$$
$$\|\mathbf{w}\| = \frac{300\sin\beta}{\sin 54^\circ}$$
$$\|\mathbf{w}\| \approx 342.37$$

Therefore, the plane must be flying at a speed of around 342.37 miles per hour with a bearing of approximately 27.41°.

Example 4: A force of magnitude 11 pounds and a force of magnitude 6 pounds act on an object at an angle of 41° with respect to one another. What is the magnitude of the resultant force, and what angle does the resultant force form with the 11-pound force?

Let **u** be the unit vector in the direction of the magnitude 11 pounds and let **v** be the unit vector in the direction of the force with magnitude 6 pounds. Let the resultant force be **w**. Then **w** = 11**u** + 6**v**.

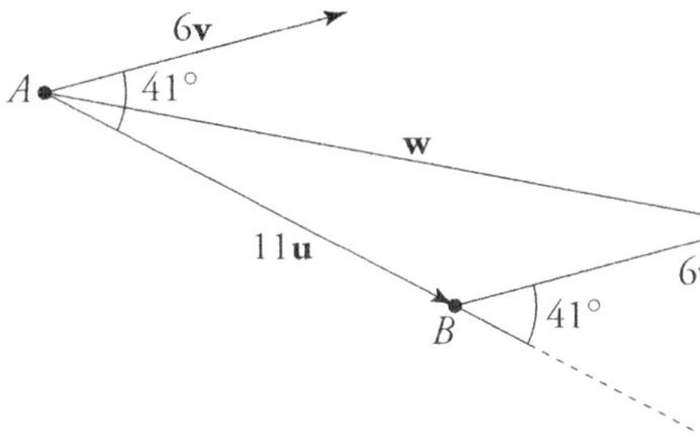

Consider ΔABC shown in the above figure; we will use the standard convention to identify the angle measures of ΔABC.

By the Law of Cosines Theorem,

$$\begin{aligned}||\mathbf{w}||^2 &= ||11\mathbf{u}||^2 + ||6\mathbf{v}||^2 - 2(||11\mathbf{u}||)(||6\mathbf{v}||)\cos\beta \\ &= 11^2 + 6^2 - 2(11)(6)\cos(180° - 41°) \\ &= 157 - 132\cos 139°\end{aligned}$$

Since $||\mathbf{w}|| > 0$, $||\mathbf{w}|| = \sqrt{157 - 132\cos 139°} \approx 16.02$.

Next, use the Law of Sines Theorem.

$$\begin{aligned}\frac{||6\mathbf{v}||}{\sin\alpha} &= \frac{||\mathbf{w}||}{\sin\beta} \\ \frac{6}{\sin\alpha} &= \frac{||\mathbf{w}||}{\sin 139°} \\ \sin\alpha &= \frac{6\sin 139°}{||\mathbf{w}||}\end{aligned}$$

Since α is a measure of an angle of a triangle, α is in the interval (0°, 180°).

There are two solutions in this interval to the basic sine equation above: $\alpha = \sin^{-1}\left(\frac{6\sin 139°}{||\mathbf{w}||}\right)$ or $\alpha = 180° - \sin^{-1}\left(\frac{6\sin 139°}{||\mathbf{w}||}\right)$. That is, α is approximately 14.22° or 165.78°. However, $\alpha < 139°$ by *Theorem 6.4*. Therefore, $\alpha \approx 14.22°$. Thus, the resultant force is 16.02 pounds, and this force makes an angle of 14.24° with the 11-pound force.

Vectors in a Rectangular Coordinate System

Introduce an *xy*-coordinate system to the plane. Let O be the origin. Let **u** be any given vector. Move **u** so that its initial point is O. Then we say the vector **u** is in *standard position*. Let $P(u_1, u_2)$ be the terminal point of **u**. We define the vector **u** (algebraically) using the coordinates of P. We use the following notation.

$$\mathbf{u} = \langle u_1, u_2 \rangle$$

The vector **u** is also called the **position vector** of point P. The number u_1 is called the *x*-component of **u**, and the number u_2 is called the *y*-component of **u**.

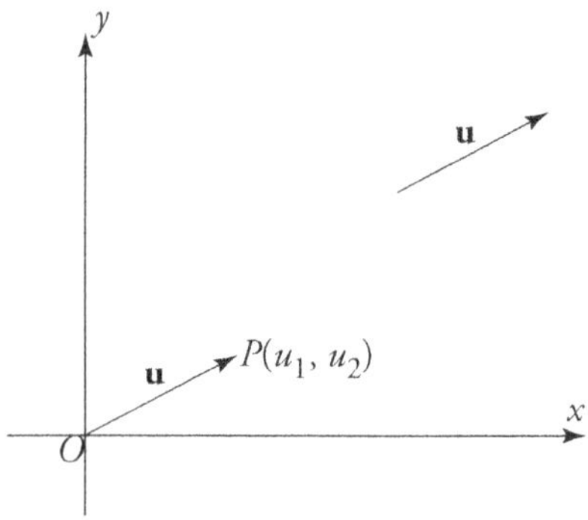

By the algebraic definition of a vector, the zero vector is $\mathbf{0} = \langle 0, 0 \rangle$.

We will define two special unit vectors: $\mathbf{i} = \langle 1, 0 \rangle$ and $\mathbf{j} = \langle 0, 1 \rangle$.

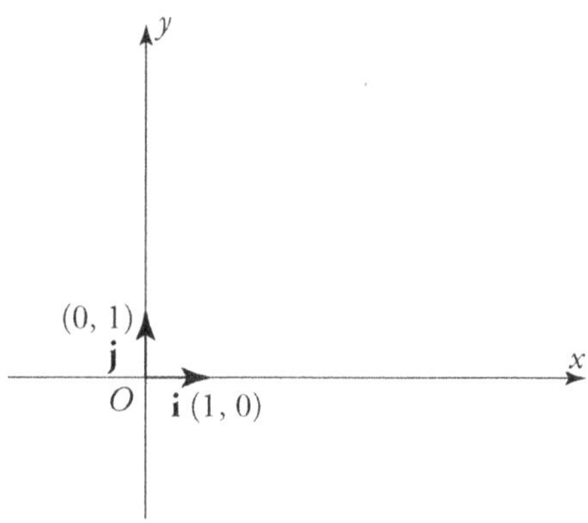

The following theorem follows from the algebraic definition of a vector.

Theorem 8.6: Suppose $\mathbf{u} = \langle u_1, u_2 \rangle$ and $\mathbf{v} = \langle v_1, v_2 \rangle$ are two vectors. Then $\mathbf{u} = \mathbf{v}$ if and only if $u_1 = v_1$ and $u_2 = v_2$.

Theorem 8.7: Let $\mathbf{u} = \langle u_1, u_2 \rangle$ be a vector in standard position. Then $\mathbf{u} = u_1\mathbf{i} + u_2\mathbf{j}$. The vector $u_1\mathbf{i}$ is the vector component of $\mathbf{u}$ in the direction of $\mathbf{i}$ and $u_2\mathbf{j}$ is the vector component of $\mathbf{u}$ in the direction of $\mathbf{j}$.

Proof: Let P be the terminal point of $\mathbf{u}$. Then point P has coordinates (u_1, u_2). We will prove this theorem if P is in quadrant 1. Proofs of other cases are left as exercises for you to perform if you'd like. Draw a perpendicular from point P to the x-axis, and let the foot of the perpendicular be point Q. Then the vector $\overrightarrow{OQ}$ has magnitude u_1, and the direction of $\overrightarrow{OQ}$ is $\mathbf{i}$. Therefore, $\overrightarrow{OQ} = u_1\mathbf{i}$ by *Theorem 8.4*. The vector $\overrightarrow{QP}$ has magnitude u_2, and the direction of $\overrightarrow{QP}$ is $\mathbf{j}$. Therefore, $\overrightarrow{QP} = u_2\mathbf{j}$ by *Theorem 8.4*. By the definition of vector addition, $\mathbf{u} = u_1\mathbf{i} + u_2\mathbf{j}$.

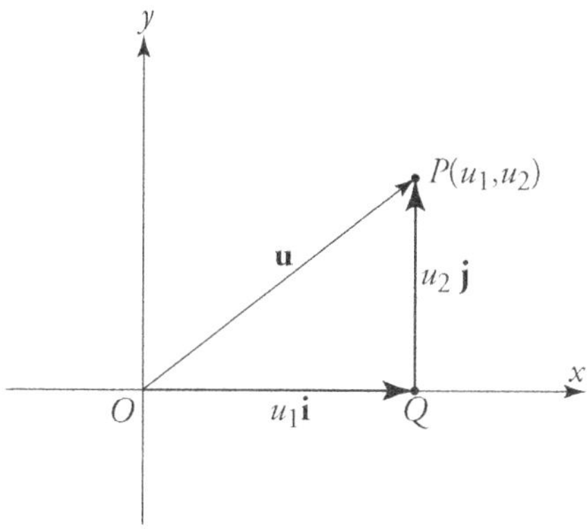

Example 5: Sketch the vector $\mathbf{u} = -2\mathbf{i} + 3\mathbf{j}$.

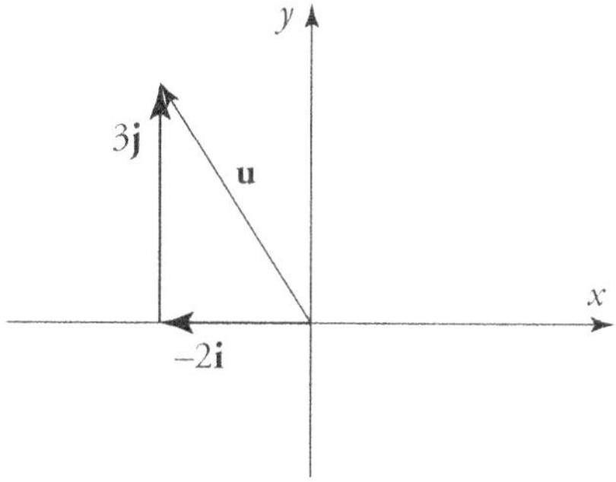

Theorem 8.8: If $\mathbf{u} = \langle u_1, u_2 \rangle$, then $\|\mathbf{u}\| = \sqrt{u_1^{\,2} + u_2^{\,2}}$.

Proof:

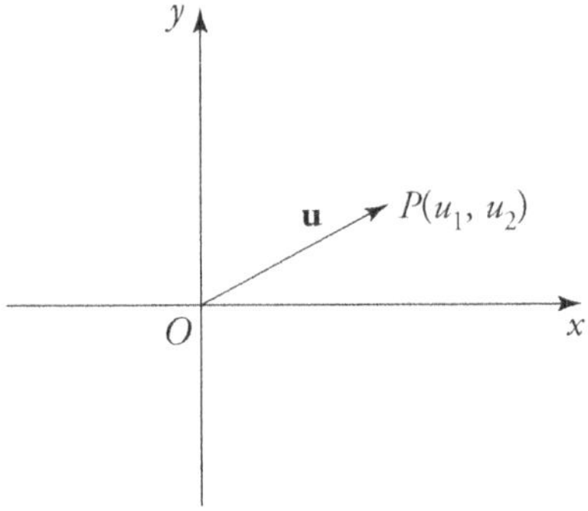

Let the terminal point of $\mathbf{u}$ be P. Then $\|\mathbf{u}\|$ is the distance between points O and P. By the two-dimensional distance formula,

$$\begin{aligned}\|\mathbf{u}\| &= \sqrt{(u_1 - 0)^2 + (u_2 - 0)^2} \\ &= \sqrt{u_1^{\,2} + u_2^{\,2}}\end{aligned}$$

Example 6: What is the magnitude of vector $\mathbf{u} = \langle 3, -5 \rangle$?

$$\begin{aligned}\|\mathbf{u}\| &= \sqrt{3^2 + (-5)^2} \\ &= \sqrt{9 + 25} \\ &= \sqrt{34}\end{aligned}$$

Example 7: Find the unit vector in the direction of the vector $\mathbf{u} = -2\mathbf{i} + 3\mathbf{j}$.

By *Theorem 8.8*, $\|\mathbf{u}\| = \sqrt{(-2)^2 + 3^2} = \sqrt{13}$. Then by *Theorem 8.3*, the unit vector in the direction of $\mathbf{u}$ is $\frac{1}{\sqrt{13}}\langle -2, 3 \rangle$.

Theorem 8.9: Let $\mathbf{u} = \langle u_1, u_2 \rangle$ and $\mathbf{v} = \langle v_1, v_2 \rangle$. Then, $\mathbf{u} + \mathbf{v} = \langle u_1 + v_1, u_2 + v_2 \rangle$.

Proof:

$$\mathbf{u} = u_1\mathbf{i} + u_2\mathbf{j} \text{ and } \mathbf{v} = v_1\mathbf{i} + v_2\mathbf{j}$$

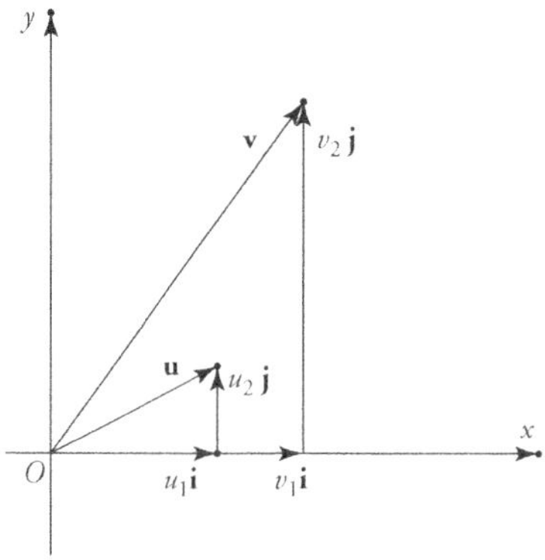

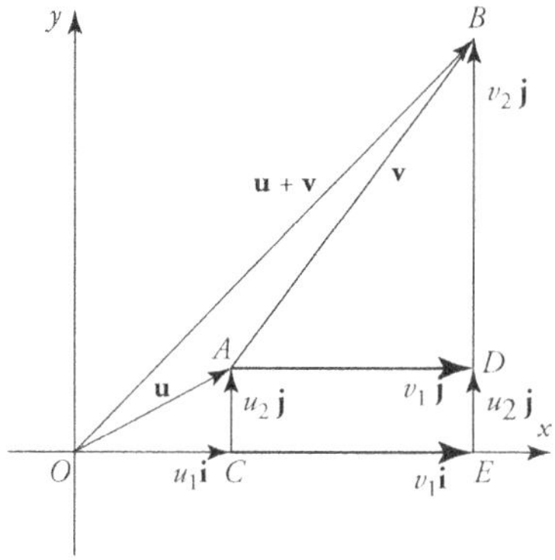

We will prove the theorem for the case where components of both vectors are positive, but the proofs of all other cases are similar. The vector **u** + **v** is obtained by the definition of vector addition. Then the vector **u** + **v** is also equal to $\overrightarrow{OE}+\overrightarrow{EB}$. The vector $\overrightarrow{OE}$ is equal to $\overrightarrow{OC}+\overrightarrow{CE}$, again by the definition of the sum of vectors. But $\overrightarrow{CE}=\overrightarrow{AD}$ (same magnitude and the same direction). The length of $OE = u_1 + v_1$ and the direction of $\overrightarrow{OE}=\mathbf{i}$. Therefore, $\overrightarrow{OE}=\left(u_1+v_1\right)\mathbf{i}$. Similarly, $\overrightarrow{EB}=\left(u_2+v_2\right)\mathbf{j}$. Then, $\mathbf{u} + \mathbf{v} = (u_1 + v_1)\mathbf{i} + (u_2 + v_2)\mathbf{j}$.

According to *Theorem 8.9*, the *x*-component of the sum vector is the sum of the *x*-components of the vectors being added, and the *y*-component of the sum vector is the sum of the *y*-components of the vectors being added.

Theorem 8.10: Let k be a scalar and $\mathbf{u} = \langle u_1, u_2\rangle$ be vector. Then the vector $k\mathbf{u} = \langle ku_1, ku_2\rangle$.

Proof: If **u** is the zero vector, then the proof is trivial. Therefore, assume that **u** is a non-zero vector. We will prove the theorem for $k > 1$ case, but the proofs of all other cases are similar.

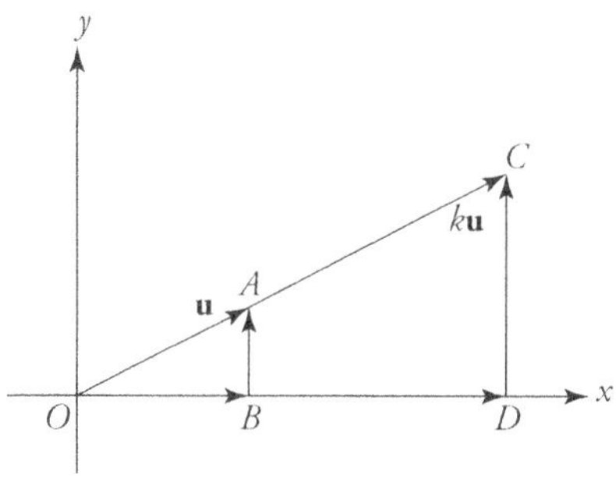

Let $\mathbf{u} = \overrightarrow{OA}$ and $k\mathbf{u} = \overrightarrow{OC}$. By the definition of scalar multiplication, the magnitude of $k\mathbf{u}$ is $k\|\mathbf{u}\|$. Therefore, $OC = k\|\mathbf{u}\|$. The right triangles ΔOAB and ΔOCD are similar by the AA Criterion Theorem. Therefore,

$$\frac{OD}{OB} = \frac{OC}{OA}$$
$$\frac{OD}{u_1} = \frac{k\|u\|}{\|u\|}$$
$$OD = ku_1$$

Similarly,

$$\frac{CD}{AB} = \frac{OC}{OA}$$
$$\frac{CD}{u_2} = \frac{k\|u\|}{\|u\|}$$
$$CD = ku_2$$

Therefore, the coordinates of point C are (ku_1, ku_2) and $k\mathbf{u} = \langle ku_1, ku_2\rangle$.

Theorem 8.11: Let $\mathbf{u} = \langle u_1, u_2\rangle$ and $\mathbf{v} = \langle v_1, v_2\rangle$. Then $\mathbf{u} - \mathbf{v} = \langle u_1 - v_1, u_2 - v_2\rangle$.

Proof: The vector $\mathbf{u} - \mathbf{v}$ is same as the vector $\mathbf{u} + (-\mathbf{v})$ by definition. By definition, vector $-\mathbf{v}$ is the vector $(-1)\mathbf{v}$. Now by *Theorem 8.10*, $(-1)\mathbf{v} = \langle -v_1, -v_2\rangle$.

Then by *Theorem 8.9*, $\mathbf{u} - \mathbf{v} = \mathbf{u} + (-1)\mathbf{v} = \langle u_1 - v_1, u_2 - v_2\rangle$.

Example 8: If the endpoints of a vector $\overrightarrow{AB}$ are $A(-2, -7)$ and $B(3, 2)$, find $\overrightarrow{AB}$.

Let $\overrightarrow{AB} = \mathbf{u}$. Let the position vector to A be $\mathbf{v}$ and the position vector to B be $\mathbf{w}$. Then $\mathbf{v} = \langle -2, -7\rangle$ and $\mathbf{w} = \langle 3, 2\rangle$.

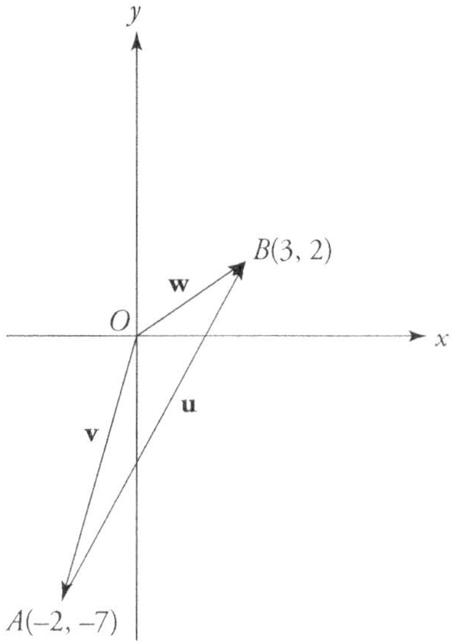

By the definition of vector addition, **u** = **w** – **v**. Then by *Theorem 8.11*,

$$\begin{aligned}\mathbf{u} &= \langle 3-(-2),\ 2-(-7)\rangle \\ &= \langle 5,\ 9\rangle\end{aligned}$$

Example 9: If **v** = ⟨8, –2⟩ and **w** = ⟨3, 7⟩, find 5**v** – 2**w**.

$$\begin{aligned}5\mathbf{v} &= \langle 5(8),\ 5(-2)\rangle \\ &= \langle 40,\ -10\rangle \text{ by } \textit{Theorem 8.10}\end{aligned}$$

Similarly,

$$\begin{aligned}2\mathbf{w} &= \langle 2(3),\ 2(7)\rangle \\ &= \langle 6,\ 14\rangle \text{ by } \textit{Theorem 8.10}\end{aligned}$$

Then by *Theorem 8.11*,

$$\begin{aligned}5\mathbf{v}-2\mathbf{w} &= \langle 40-6,\ -10-14\rangle \\ &= \langle 34,\ -24\rangle\end{aligned}$$

Example 10: Find a unit vector **v** with the same direction as the vector **u**, given that **u** = ⟨7, –1⟩.

By *Theorem 8.8*,

$$\begin{aligned}\|\mathbf{u}\| &= \sqrt{7^2 + (-1)^2}\\ &= \sqrt{49+1}\\ &= 5\sqrt{2}\end{aligned}$$

Then by *Theorem 8.3*,

$$\frac{1}{\|\mathbf{u}\|}\mathbf{u} = \frac{1}{5\sqrt{2}}\langle 7, -1\rangle$$

Finally, by *Theorem 8.10*,

$$\begin{aligned}\frac{1}{\|\mathbf{u}\|}\mathbf{u} &= \left\langle \frac{7}{5\sqrt{2}}, \frac{-1}{5\sqrt{2}} \right\rangle\\ &= \left\langle \frac{7\sqrt{2}}{10}, \frac{-\sqrt{2}}{10} \right\rangle\end{aligned}$$

We already know that the commutative property holds for vectors. (See *Theorem 8.1.*) Example 11 shows an algebraic proof of *Theorem 8.1.*

Example 11: If **u** and **v** are vectors, show that **u** + **v** = **v** + **u**.

Let **u** = $\langle u_1, u_2\rangle$ and **v** = $\langle v_1, v_2\rangle$. Then by *Theorem 8.9*, **u** + **v** = $\langle u_1 + v_1, u_2 + v_2\rangle$.

Also, by *Theorem 8.9*, **v** + **u** = $\langle v_1 + u_1, v_2 + u_2\rangle$.

Since real numbers are commutative, $v_1 + u_1 = u_1 + v_1$ and $v_2 + u_2 = u_2 + v_2$. Therefore, **u** + **v** = **v** + **u**.

The following theorems follow easily from *Theorems 8.9* and *8.10.*

Theorem 8.12: If **u**, **v**, and **w** are vectors, then **u** + (**v** + **w**) = (**u** + **v**) + **w**.

Theorem 8.13: If **u** and **v** are vectors and k is a scalar, then k(**u** + **v**) = k**u** + k**v**.

Theorem 8.14: If m and n are scalars and **v** is a vector, then $(m + n)$**v** = m**v** + n**v**.

Theorem 8.15: If **v** is a vector, then **v** + **0** = **0** + **v** = **v**.

Theorem 8.16: If **v** is a vector, then 1**v** = **v**.

Theorem 8.17: If **v** is a vector, then **v** + (−**v**) = **0**.

The Dot Product of Two Vectors

Given two vectors, $\mathbf{u} = \langle u_1, u_2 \rangle$ and $\mathbf{v} = \langle v_1, v_2 \rangle$, we define the **dot product,** between **u** and **v**, denoted by **u** · **v**, and pronounced "u dot v," as:

$$\mathbf{u} \cdot \mathbf{v} = u_1 v_1 + u_2 v_2$$

The quantity **u** · **v** is a scalar, by the definition.

The following theorem follows directly from the definition of the dot product.

Theorem 8.18: If **u** and **v** are vectors, then **u** · **v** = **v** · **u**.

The following theorem follows from *Theorem 8.9* and the definition of the dot product.

Theorem 8.19: If **u**, **v**, and **w** are vectors, then **u** · (**v** + **w**) = **u** · **v** + **u** · **w**.

The following theorem follows from the definition of the dot product and *Theorem 8.9.*

Theorem 8.20: If **u** and **v** are vectors and k is a scalar, then $\mathbf{u} \cdot (k\mathbf{v}) = (k\mathbf{u}) \cdot \mathbf{v} = k(\mathbf{u} \cdot \mathbf{v})$.

Theorem 8.21: If **u** is a vector, then $\mathbf{u} \cdot \mathbf{u} = \|\mathbf{u}\|^2$.

Proof: Let $\mathbf{u} = \langle u_1, u_2 \rangle$. Then by the definition of the dot product,

$$\begin{aligned} \mathbf{u} \cdot \mathbf{u} &= u_1 u_1 + u_2 u_2 \\ &= u_1^2 + u_2^2 \\ &= \left(\sqrt{u_1^2 + u_2^2}\right)^2 \\ &= \|\mathbf{u}\|^2 \end{aligned}$$

Theorem 8.22: If **u** and **v** are vectors and θ is the (measure of the) angle between them, then $\mathbf{u} \cdot \mathbf{v} = \|\mathbf{u}\| \, \|\mathbf{v}\| \cos \theta$.

Proof: If at least one of **u** or **v** is the zero vector, then it is easy to show that the theorem is true. Therefore, assume that both **u** and **v** are non-zero.

Recall that $0 \le \theta \le \pi$. Suppose $\theta = 0$. Then **u** and **v** have the same direction and $\mathbf{v} = k\mathbf{u}$ for some positive scalar k.

The left side of the theorem is $\mathbf{u} \cdot \mathbf{v} = \mathbf{u} \cdot (k\mathbf{u}) = k(\mathbf{u} \cdot \mathbf{u})$, by *Theorem 8.20.*

Then by *Theorem 8.21*, the left side of the theorem is $k\|\mathbf{u}\|^2$.

The right side of the theorem is,

$$\begin{aligned} \|\mathbf{u}\|\|\mathbf{v}\|\cos 0 &= \|\mathbf{u}\|\|k\mathbf{u}\| \\ &= k\|\mathbf{u}\|^2 \end{aligned}$$

The last equality follows from the definition of scalar multiplication.

Now, suppose $\theta = \pi$. Then **u** and **v** have opposite directions, and $\mathbf{v} = k\mathbf{u}$ for some negative scalar k.

The left side of the theorem is $\mathbf{u} \cdot \mathbf{v} = \mathbf{u} \cdot (k\mathbf{u}) = k(\mathbf{u} \cdot \mathbf{u})$ by *Theorem 8.20.*

Then by *Theorem 8.21*, the left side of the theorem is $k\|\mathbf{u}\|^2$.

The right side of the theorem is,

$$\begin{aligned} \|\mathbf{u}\|\|\mathbf{v}\|\cos \pi &= -\|\mathbf{u}\|\|k\mathbf{u}\| \\ &= -(-k)\|\mathbf{u}\|^2 \\ &= k\|\mathbf{u}\|^2 \end{aligned}$$

The next-to-last equality follows from the definition of scalar multiplication.

Now, suppose $0 < \theta < \pi$.

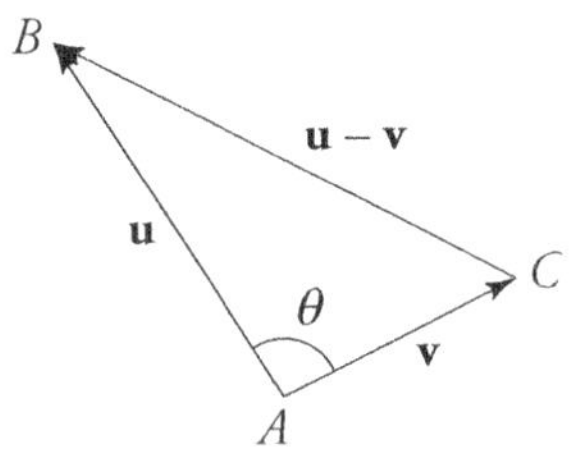

Let $\mathbf{u} = \overrightarrow{AB}$ and $\mathbf{v} = \overrightarrow{AC}$, as shown in the figure above. Then by the definition of vector addition, $\overrightarrow{CB} = \mathbf{u} - \mathbf{v}$. By using the Law of Cosines Theorem for ΔABC, $CB^2 = AB^2 + AC^2 - 2(AB)(AC)\cos\theta$. That is,

$$\|\mathbf{u} - \mathbf{v}\|^2 = \|\mathbf{u}\|^2 + \|\mathbf{v}\|^2 - 2\|\mathbf{u}\|\,\|\mathbf{v}\|\cos\theta \qquad (1)$$

By *Theorem 8.21*, $\|\mathbf{u}-\mathbf{v}\|^2 = (\mathbf{u}-\mathbf{v}) \cdot (\mathbf{u}-\mathbf{v})$. Therefore,

$$\begin{aligned}\text{left side of (1)} &= (\mathbf{u}-\mathbf{v})\cdot(\mathbf{u}-\mathbf{v}) \\ &= \mathbf{u}\cdot\mathbf{u} - 2\mathbf{u}\cdot\mathbf{v} + \mathbf{v}\cdot\mathbf{v}\end{aligned}$$

The last equality follows from *Theorems 8.19, 8.18,* and *8.20.* Then by *Theorem 8.21*,

$$\text{left side of (1)} = \|\mathbf{u}\|^2 - 2\mathbf{u}\cdot\mathbf{v} + \|\mathbf{v}\|^2$$

Substitute in (1):

$$\|\mathbf{u}\|^2 - 2\mathbf{u}\cdot\mathbf{v} + \|\mathbf{v}\|^2 = \|\mathbf{u}\|^2 + \|\mathbf{v}\|^2 - 2\|\mathbf{u}\|\,\|\mathbf{v}\| \cos\theta$$

By combining like terms and simplifying, we get $\mathbf{u}\cdot\mathbf{v} = \|\mathbf{u}\|\,\|\mathbf{v}\|\cos\theta$.

The following important theorem is a corollary of *Theorem 8.22* (i.e., the proof follows from *Theorem 8.22*).

Theorem 8.23: If **u** and **v** are two non-zero vectors, then
(1) The angle between **u** and **v** is acute if and only if $\mathbf{u}\cdot\mathbf{v} > 0$.
(2) The angle between **u** and **v** is obtuse if and only if $\mathbf{u}\cdot\mathbf{v} < 0$.
(3) **u** and **v** are orthogonal if and only if $\mathbf{u}\cdot\mathbf{v} = 0$.

Proof: We will prove part (3). The proofs of the other two parts are left as exercises for you to perform if you'd like.

Suppose **u** and **v** are orthogonal. That is, the angle between them is $\frac{\pi}{2}$.
Then by *Theorem 8.22*, $\mathbf{u}\cdot\mathbf{v} = \|\mathbf{u}\|\|\mathbf{v}\|\cos\frac{\pi}{2} = 0$.
Now suppose $\mathbf{u}\cdot\mathbf{v} = 0$. Then by *Theorem 8.22*, $\|\mathbf{u}\|\,\|\mathbf{v}\|\cos\theta = 0$, where $0 \le \theta \le \pi$. Since **u** and **v** are two non-zero vectors, $\|\mathbf{u}\| \ne 0$ and $\|\mathbf{v}\| \ne 0$. Therefore, $\cos\theta = 0$.

The only solution of the above basic cosine equation in the interval $[0, \pi]$ is $\frac{\pi}{2}$. Therefore, **u** and **v** are orthogonal.

Example 12: Given that $\mathbf{u} = \langle 5, -3\rangle$ and $\mathbf{v} = \langle 6, 10\rangle$, show that **u** and **v** are orthogonal.

By the definition of the dot product,

$$\begin{aligned}\mathbf{u}\cdot\mathbf{v} &= 5(6)+(-3)(10) \\ &= 0\end{aligned}$$

Therefore, **u** and **v** are orthogonal by *Theorem 8.23*.

Example 13: What is the measure of the angle between **u** = ⟨5, –2⟩ and **v** = ⟨6, 11⟩?

By the definition of the dot product,

$$\begin{aligned}\mathbf{u}\cdot\mathbf{v} &= (5)(6)+(-2)(11)\\ &= 8\end{aligned}$$

We know that the angle between **u** and **v** is acute by *Theorem 8.23.*

$$\|\mathbf{u}\| = \sqrt{5^2+(2)^2} = \sqrt{29} \text{ and } \|\mathbf{v}\| = \sqrt{6^2+11^2} = \sqrt{157}$$

By *Theorem 8.22,* $\mathbf{u}\cdot\mathbf{v} = \|\mathbf{u}\|\,\|\mathbf{v}\|\cos\theta$, where θ is the measure of the angle between **u** and **v**. So, $8 = \sqrt{29}\sqrt{157}\cos\theta$ and $\cos\theta = \frac{8}{\sqrt{4553}}$.

Since we know the angle is acute, $\theta = \cos^{-1}\left(\frac{8}{\sqrt{4553}}\right) \approx 83.19°$.

Theorem 8.24: Let **u** be a non-zero vector in standard position and let θ be the measure of the angle between **u** and **i**. Then $\mathbf{u} = \langle \|\mathbf{u}\|\cos\theta, \|\mathbf{u}\|\sin\theta\rangle$.

Proof: The angle θ is in standard position. Let $P(x, y)$ be the endpoint of **u**. Then point P is on the terminal side of θ, and by the definitions of sin θ and cos θ, $\cos\theta = \frac{x}{\|\mathbf{u}\|}$ and $\sin\theta = \frac{y}{\|\mathbf{u}\|}$.

Therefore, $x = \|\mathbf{u}\|\cos\theta$, $y = \|\mathbf{u}\|\sin\theta$, and by the algebraic definition of **u**, $\mathbf{u} = \langle \|\mathbf{u}\|\cos\theta, \|\mathbf{u}\|\sin\theta\rangle$.

An object is said to be in a state of **static equilibrium** if all the forces acting on the object add up to zero.

Example 14: A tightrope walker weighing 150 pounds is standing closer to one end of the rope than the other. The shorter length of rope deflects 5° from the horizontal. The longer length of rope deflects 3°. What is the tension on each part of the rope?

Let the tension (vector) on the short end of the rope be **u**, let the tension on the long end of the rope be **v**, and let the weight (vector) of the tightrope walker be **w**. The following is the force diagram with all three force vectors in standard position.

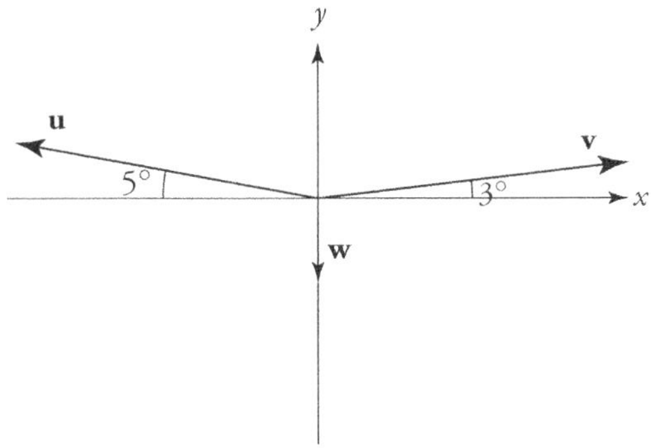

By *Theorem 8.24*,

$\mathbf{u} = \langle \|\mathbf{u}\| \cos 175°, \|\mathbf{u}\| \sin 175° \rangle = \langle -\|\mathbf{u}\| \cos 5°, \|\mathbf{u}\| \sin 5° \rangle$,

$\mathbf{v} = \langle \|\mathbf{v}\| \cos 3°, \|\mathbf{v}\| \sin 3° \rangle$, and

$\mathbf{w} = \langle \|\mathbf{w}\| \cos 270°, \|\mathbf{w}\| \sin 270° \rangle = \langle 0, -\|\mathbf{w}\| \rangle$.

Since the system is in static equilibrium, $\mathbf{u} + \mathbf{v} + \mathbf{w} = 0$ or $\mathbf{u} + \mathbf{v} = -\mathbf{w}$.

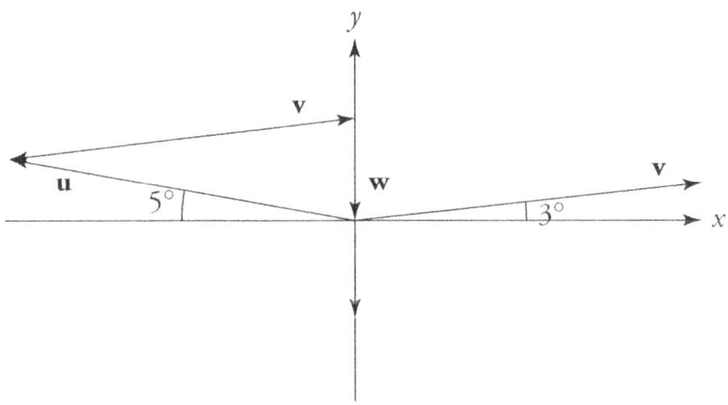

Therefore,

$$\langle -\|\mathbf{u}\| \cos 5°, \|\mathbf{u}\| \sin 5° \rangle + \langle \|\mathbf{v}\| \cos 3°, \|\mathbf{v}\| \sin 3° \rangle = \langle 0, \|\mathbf{w}\| \rangle$$

$$\langle -\|\mathbf{u}\| \cos 5° + \|\mathbf{v}\| \cos 3°, \|\mathbf{u}\| \sin 5° + \|\mathbf{v}\| \sin 3° \rangle = \langle 0, 150 \rangle$$

Then from *Theorem 8.6*,

$$-\|\mathbf{u}\|\cos 5^\circ + \|\mathbf{v}\|\cos 3^\circ = 0 \qquad (1)$$
$$\|\mathbf{u}\|\sin 5^\circ + \|\mathbf{v}\|\sin 3^\circ = 150 \qquad (2)$$

Multiply (1) by sin 3° and multiply (2) by cos 3°.

$$-\|\mathbf{u}\|\cos 5^\circ \sin 3^\circ + \|\mathbf{v}\|\cos 3^\circ \sin 3^\circ = 0 \qquad (3)$$
$$\|\mathbf{u}\|\sin 5^\circ \cos 3^\circ + \|\mathbf{v}\|\sin 3^\circ \cos 3^\circ = 150 \cos 3^\circ \qquad (4)$$

Subtract (3) from (4):

$$\|\mathbf{u}\| \sin 5^\circ \cos 3^\circ + \|\mathbf{u}\| \cos 5^\circ \sin 3^\circ = 150 \cos 3^\circ$$

Solve this equation for $\|\mathbf{u}\|$ and we get:

$$\|\mathbf{u}\| = \frac{150\cos 3^\circ}{(\sin 5^\circ \cos 3^\circ + \cos 5^\circ \sin 3^\circ)} = \frac{150\cos 3^\circ}{\sin 8^\circ}$$

Multiply (1) by sin 5° and multiply (2) by cos 5°.

$$-\|\mathbf{u}\|\cos 5^\circ \sin 5^\circ + \|\mathbf{v}\|\cos 3^\circ \sin 5^\circ = 0 \qquad (5)$$
$$\|\mathbf{u}\|\sin 5^\circ \cos 5^\circ + \|\mathbf{v}\|\sin 3^\circ \cos 5^\circ = 150 \cos 5^\circ \qquad (6)$$

Add (5) and (6):

$$\|\mathbf{v}\| \cos 3^\circ \sin 5^\circ + \|\mathbf{v}\| \sin 3^\circ \cos 5^\circ = 150 \cos 5^\circ$$

Solve this equation for $\|\mathbf{v}\|$ and we get:

$$\|\mathbf{v}\| = \frac{150\cos 5^\circ}{(\cos 3^\circ \sin 5^\circ + \sin 3^\circ \cos 5^\circ)} = \frac{150\cos 5^\circ}{\sin 8^\circ}$$

Using a calculator, $\|\mathbf{u}\| \approx$ 1,076 lbs. and $\|\mathbf{v}\| \approx$ 1,074 lbs.

Chapter Check-Out

Questions

1. An airplane is traveling with an airspeed of 225 mph at a bearing of 205°. A 60-mph wind is blowing with a bearing of 100°. What is the resultant ground speed and direction of the plane?
2. A force of 22 pounds and a force of 35 pounds act on an object at an angle of 32° with respect to one another. What is the resultant force on the object?
3. If the endpoints of a vector $\overrightarrow{AB}$ have coordinates of $A(-4, 6)$ and $B(10, 4)$, what is $\overrightarrow{AB}$?
4. True or False: The dot product of two orthogonal vectors is always zero.
5. What is the measure of the angle between vector $\mathbf{u} = \langle 5, 7 \rangle$ and $\mathbf{v} = \langle -6, 6 \rangle$?

Answers

1. 217 mph, bearing 189.53°
2. 54.91 pounds
3. $\langle 14, -2 \rangle$
4. True
5. 80.54°

Chapter 9

POLAR COORDINATES AND COMPLEX NUMBERS

Chapter Check-In

- ❑ Defining polar coordinates
- ❑ Recognizing the graphs of some basic polar curves
- ❑ Converting between polar and rectangular coordinates
- ❑ Defining complex numbers and their operations
- ❑ Multiplying and dividing complex numbers in polar form
- ❑ Defining and using De Moivre's Theorem
- ❑ Taking powers and roots of complex numbers in polar form

Common Core Standard

Please refer to the standards for mathematical practice, found here: www.corestandards.org/Math/Practice/.

We are familiar with the rectangular coordinate system of the plane. There is another coordinate system of the plane that is useful in many problem-solving situations. We need this coordinate system to study complex numbers, and this chapter is a proper place to introduce it. It is known as the **polar coordinate system.**

Trigonometry Theorems

The trigonometry theorems in this book are numbered for organizational purposes. For example, *Theorem 2.2* is the second theorem given in Chapter 2. Study these theorems by their content, not by their number, as the theorem numbering has no significance outside of this book.

Polar Coordinates

Fix a point on the plane. We will call this point the **pole** and denote it by O. Fix a ray emanating from O. We will call this ray the **polar axis.** By rotating the plane around O, we can make this axis point in any direction. For our purposes, we will assume that the polar axis is horizontal.

Now, consider any point P in the plane other than O. Let **s** be the angle with vertex O and the initial side equal to the polar axis so that point P lies either on the terminal side of θ or on the extended side of θ. If θ is measured counterclockwise from the polar axis, then we assign a positive sign to θ. If θ is measured clockwise from the polar axis, then we assign a negative sign to θ. Let r be the length of the segment $\overline{OP}$. If point P lies on the terminal side of θ, then we assign a positive sign to r. If point P lies on the extended terminal side θ, then we assign a negative sign to r.

The ordered pair (r, θ) is called the **polar coordinates** of P.

We choose polar coordinates of O to be $(0, \theta)$, where θ is any arbitrary angle.

Example 1: Plot the points with the following polar coordinates: $\left(2, \frac{\pi}{6}\right)$, $\left(2, -\frac{5\pi}{6}\right)$, $\left(-2, \frac{\pi}{6}\right)$, and $\left(-2, -\frac{5\pi}{6}\right)$.

The following figures show polar coordinates of the given points.

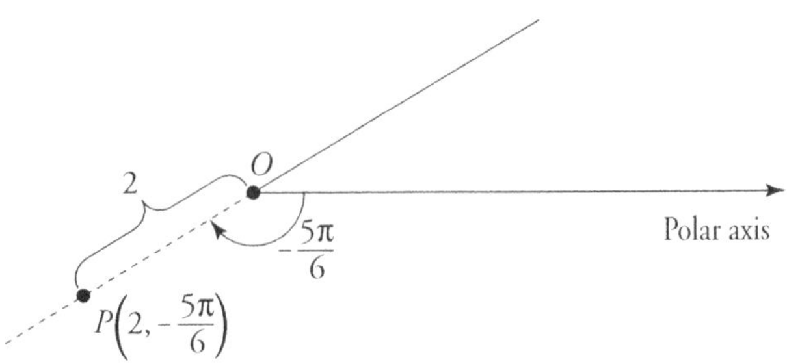

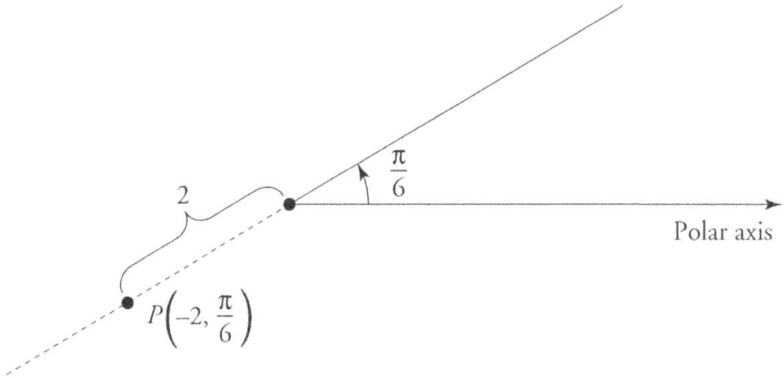

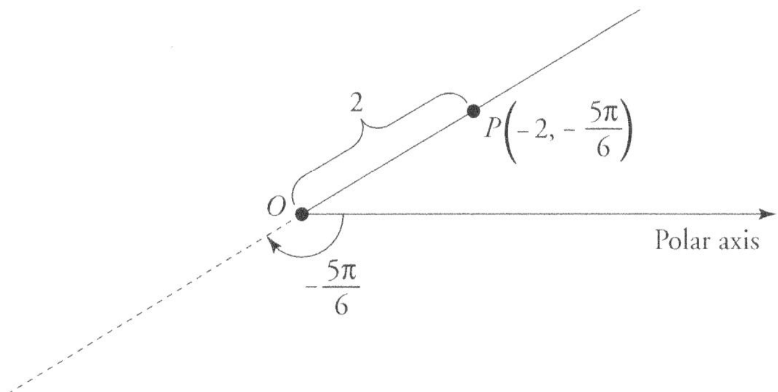

Graphs of Polar Equations

An equation in r and θ is called a *polar equation*. For example, the following are some polar equations.

$$r = 1,\ \theta = \frac{\pi}{4},\ r = 2\cos\theta,\ r = 2\sin\theta$$

The collection of all points (r, θ) for which a given polar equation is true (i.e., all solutions of the equation) is called the *graph of the polar equation*.

Example 2: Sketch the graph of $r = 1$ in a polar coordinate system.

Any point of the form $(1, \theta)$, for an arbitrary angle θ, is a solution of the equation $r = 1$. Therefore, the graph of $r = 1$ is the unit circle with center O.

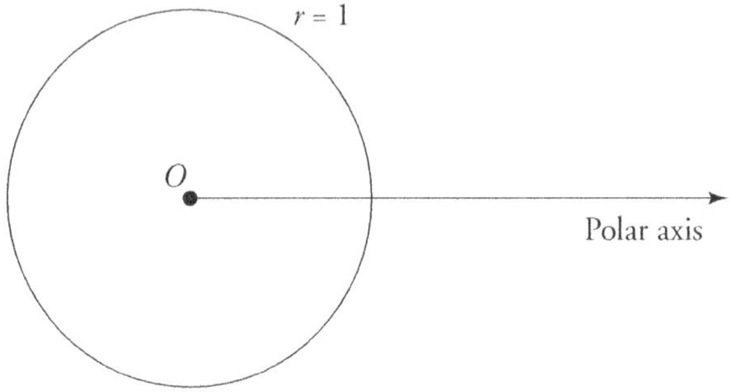

Example 3: Sketch the graph of $\theta = \frac{\pi}{4}$ in a polar coordinate system.

Any point of the form $\left(r, \frac{\pi}{4}\right)$, for an arbitrary real number r, is a solution of the equation $\theta = \frac{\pi}{4}$. Therefore, the graph of $\theta = \frac{\pi}{4}$ is the line passing through O that makes a counterclockwise angle of $\frac{\pi}{4}$ with the polar axis.

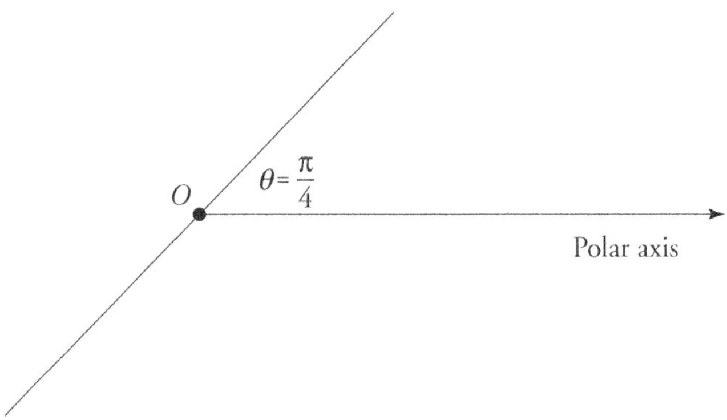

Example 4: Sketch the graph of $r = 2 \cos \theta$ in a polar coordinate system.

First, we will make a table using special angles and their multiples, and then we will plot those points on a polar coordinate system.

$r = 2\cos\theta$	θ
2	0
$\sqrt{3}$	$\frac{\pi}{6}$
$\sqrt{2}$	$\frac{\pi}{4}$
1	$\frac{\pi}{3}$
0	$\frac{\pi}{2}$

$r = 2\cos\theta$	θ
−1	$\frac{2\pi}{3}$
$-\sqrt{2}$	$\frac{3\pi}{4}$
$-\sqrt{3}$	$\frac{5\pi}{6}$
−2	π

Now, we will plot these points on a polar coordinate system.

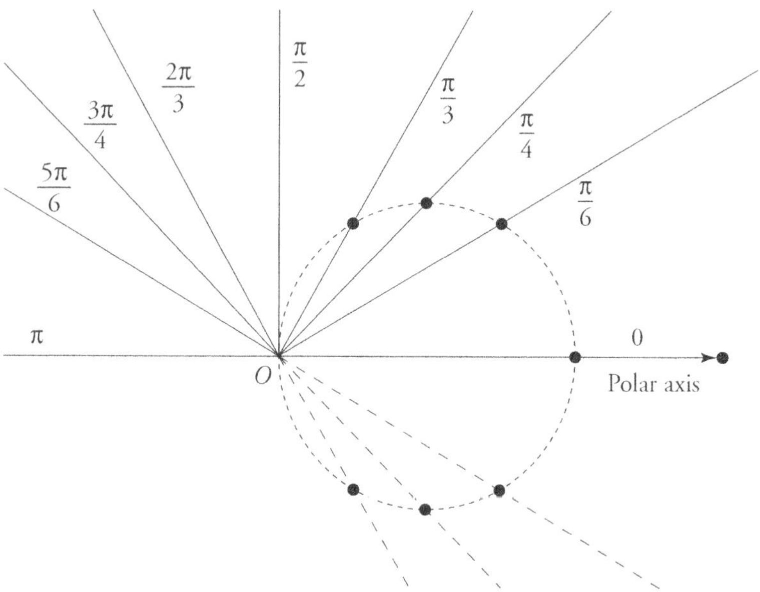

It looks like the graph of the equation $r = 2\cos\theta$ is a circle.

Relationships Between Polar and Rectangular Coordinates

Introduce an *xy*-coordinate system to the plane as follows. Choose the pole O as the origin and the positive x-axis as the polar axis. The following is the plane with x-axis, y-axis, pole, and the polar axis.

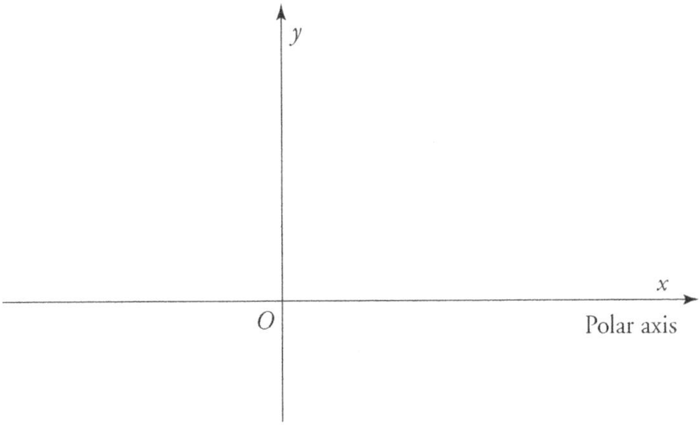

Now, any point on this plane can be represented using polar coordinates as well as rectangular coordinates. However, we have one other problem. While rectangular coordinates of a point are unique (See *CliffsNotes Geometry Common Core Quick Review*), the polar coordinates of a point are not. For example, the point with polar coordinates $\left(2,\ \frac{\pi}{6}\right)$ has other polar coordinates, such as $\left(-2,\ -\frac{5\pi}{6}\right)$, $\left(2,\ -\frac{11\pi}{6}\right)$, $\left(2,\ \frac{13\pi}{6}\right)$, etc. In fact, a given point has infinitely many polar coordinates.

We can make polar coordinates of a point unique by putting restrictions on r and θ. We require that $r > 0$ and $0 \le \theta < 2\pi$. Under these restrictions, any point on the plane has unique polar coordinates.

Consider a point P with *xy*-coordinates (x, y) and polar coordinates (r, θ), as shown in the following figure.

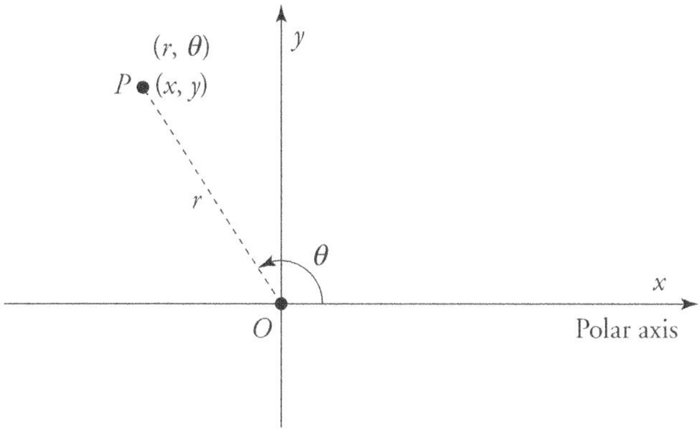

By the definitions of cos θ and sin θ, $\frac{x}{r} = \cos\theta$ and $\frac{y}{r} = \sin\theta$. Therefore, $x = r\cos\theta$ and $y = r\sin\theta$.

By dividing y by x, we get $\frac{y}{x} = \frac{r\sin\theta}{r\cos\theta} = \tan\theta$.

By the two-dimensional distance formula, $r = \sqrt{x^2 + y^2}$.

Therefore, the following are the relationships between polar coordinates and rectangular coordinates of a point in the plane.

Table 9-1 Polar and Rectangular Coordinate Relationships

$x = r\cos\theta$	$r = \sqrt{x^2 + y^2}$
$y = r\sin\theta$	$\tan\theta = \frac{y}{x}$

Example 5: Find the polar coordinates of point P if the rectangular coordinates of P are (1, 1).

Point P is in the first quadrant. Therefore, $0 < \theta < \frac{\pi}{2}$, and $\tan\theta = 1$ since $x = 1$ and $y = 1$.

The only solution of this basic tangent equation in the first quadrant is $\frac{\pi}{4}$, and $r = \sqrt{1^2 + 1^2} = \sqrt{2}$. Therefore, the polar coordinates of P are $\left(\sqrt{2}, \frac{\pi}{4}\right)$.

Example 6: Find the polar coordinates of point P if the rectangular coordinates of P are $(-1, -1)$.

Point P is in the third quadrant. Therefore, $\pi < \theta < \frac{3\pi}{2}$. Again, $\tan\theta = 1$, since $x = -1$ and $y = -1$.

The only solution of this basic tangent equation in the third quadrant is $\frac{5\pi}{4}$, and $r = \sqrt{(-1)^2 + (-1)^2} = \sqrt{2}$. Therefore, the polar coordinates of P are $\left(\sqrt{2}, \frac{5\pi}{4}\right)$.

The following is a theorem that you have learned in algebra.

Theorem 9.1: The equation of a circle with center (h, k) and radius $r > 0$ is $(x - h)^2 + (y - k)^2 = r^2$.

Proof: This theorem is proven using the two-dimensional distance formula.

Example 7: Transform the equation $r = 2\cos\theta$ to rectangular coordinates.

$$r = 2\cos\theta$$

Multiply both sides by r.

$$r^2 = 2r\cos\theta$$

$$x^2 + y^2 = 2x$$

$$x^2 - 2x + y^2 = 0$$

$$\left(x^2 - 2x + 1\right) + y^2 = 1$$

$$\left(x - 1\right)^2 + y^2 = 1$$

By *Theorem 9.1*, the graph of the above equation is the circle with radius 1 and center $(1, 0)$. Going back to Example 4, this is a confirmation that the graph we obtained in Example 4 is a circle.

Example 8: Transform the equation $x^2 + y^2 + 5x = 0$ to polar coordinate form.

$$x^2 + y^2 + 5x = 0$$

$$r^2 + 5x = 0$$

$$r^2 + 5(r\cos\theta) = 0$$

$$r^2 + 5r\cos\theta = 0$$

$$r(r + 5\cos\theta) = 0$$

The equation $r = 0$ indicates the pole. Thus, keep only the other equation, $r + 5\cos\theta = 0$.

Example 9: Use the relationships between rectangular and polar coordinates to identify and sketch the graph of $r = 2\sin\theta$ in polar coordinates.

$$r = 2\sin\theta$$
$$r^2 = 2r\sin\theta$$
$$x^2 + y^2 = 2y$$
$$x^2 + y^2 - 2y = 0$$
$$x^2 + (y^2 - 2y + 1) = 1$$
$$x^2 + (y-1)^2 = 1$$

This is the equation of a circle with center (0, 1) and radius 1 in rectangular coordinates. The following is the graph of $r = 2\sin\theta$ in polar coordinates.

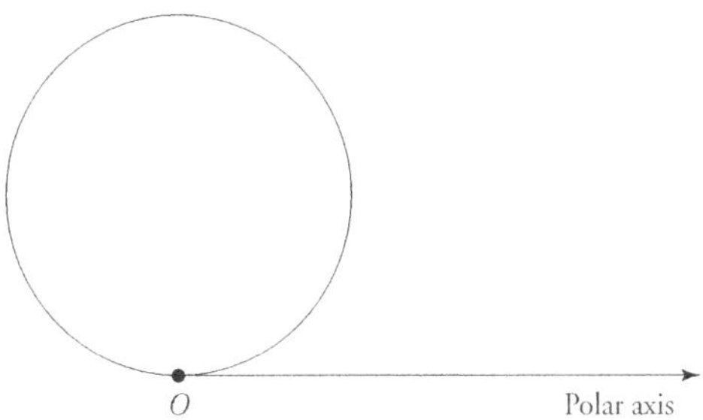

Graphs of some special polar equations are given in the following figures.

Circle: $r = a$

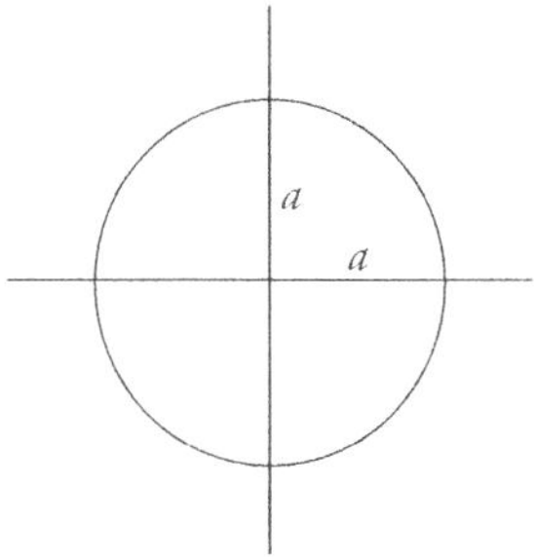

(a)

Circle: $r = a\cos\theta$

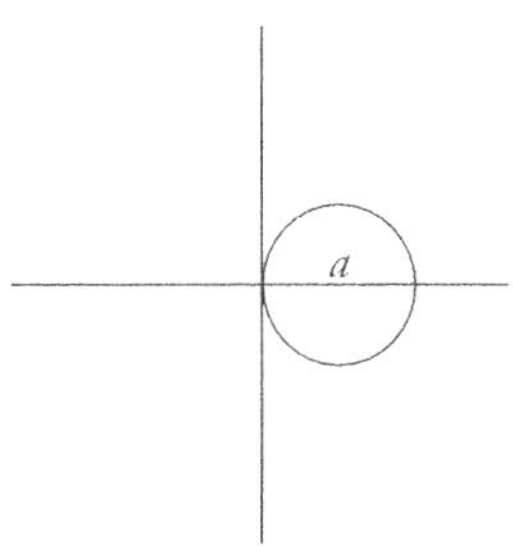

(b)

Circle: $r = a \sin \theta$

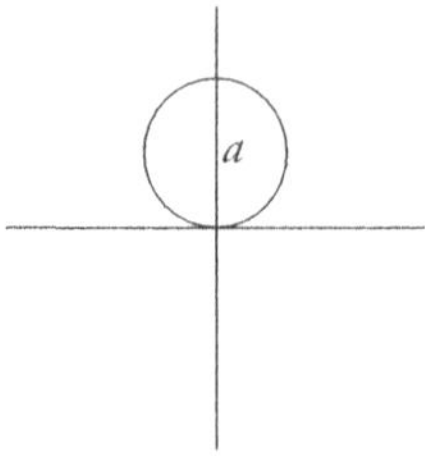

(c)

Line: $\theta = a$

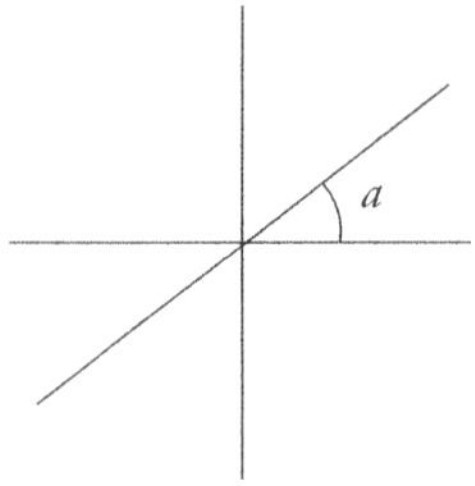

(d)

Horizontal line: $r = \dfrac{a}{\sin \theta}$

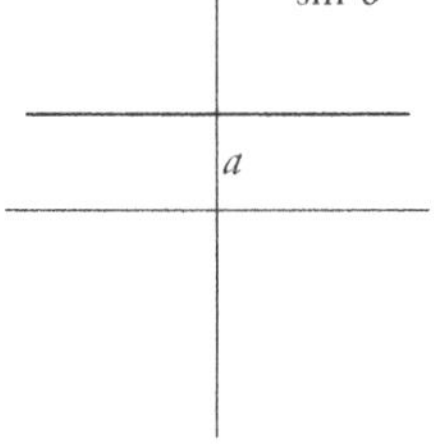

(e)

Vertical line: $r = \dfrac{a}{\cos \theta}$

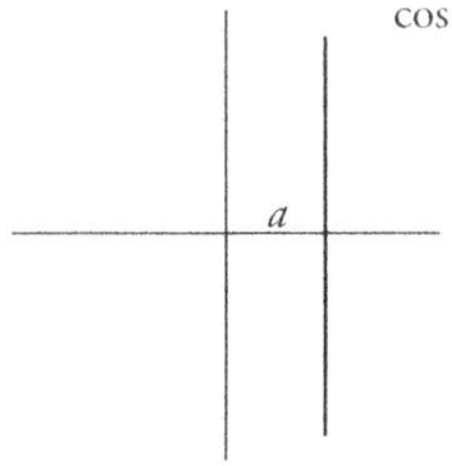

(f)

Cardioid: $r = a + a \cos \theta$

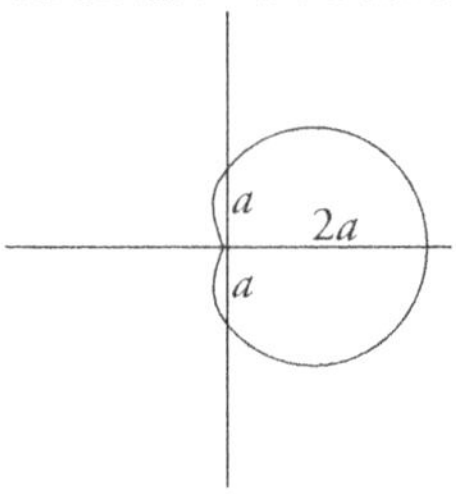

(g)

Cardioid: $r = a + a \sin \theta$

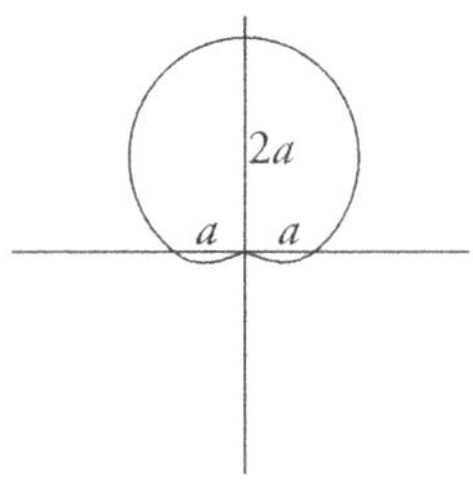

(h)

Archimedes' spiral: $r = a\,\theta$

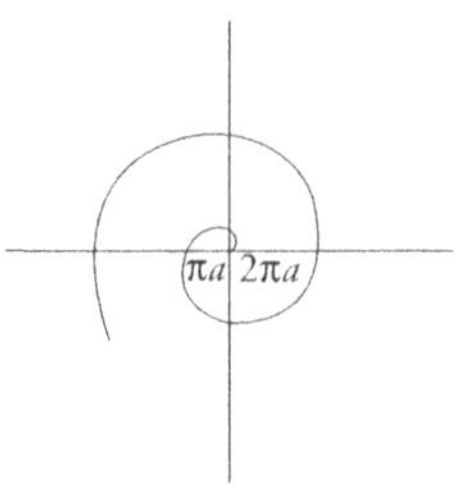

(i)

Three-leaved rose: $r = a \cos 3\,\theta$

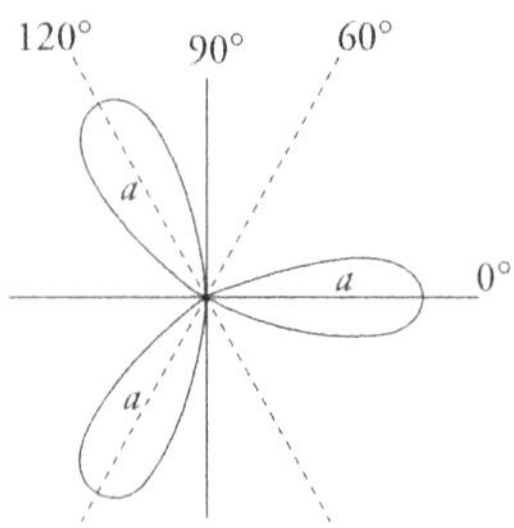

(j)

Three-leaved rose: $r = a \sin 3\theta$

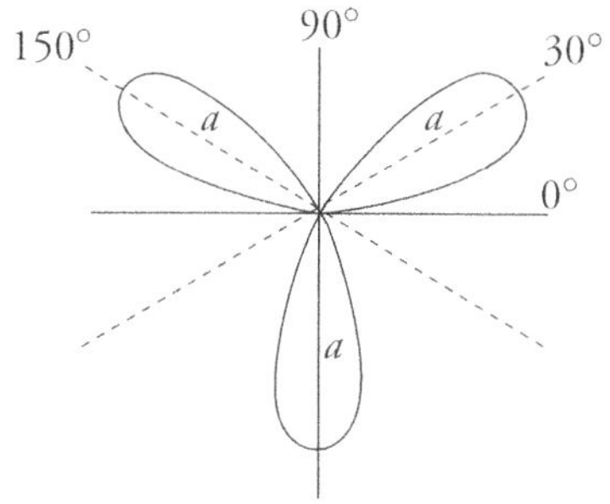

(k)

Lemniscate: $r^2 = a^2 \cos 2\theta$

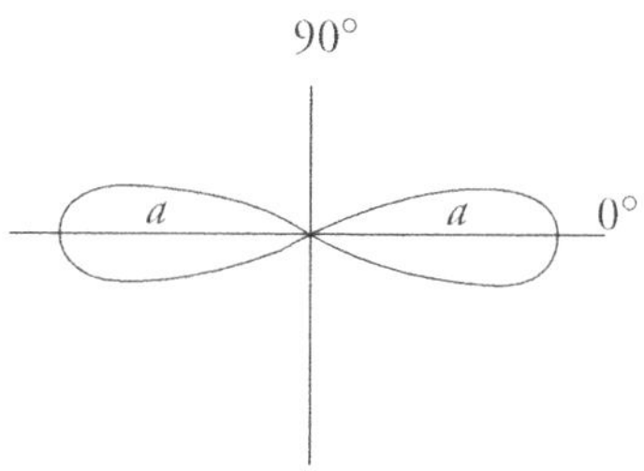

(l)

Lemniscate: $r^2 = a^2 \sin 2\theta$

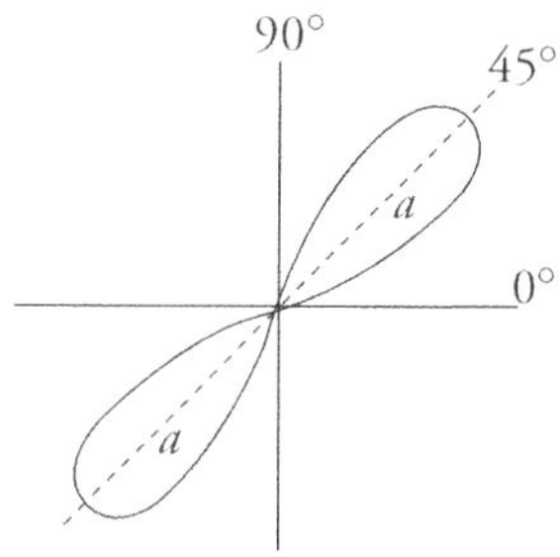

(m)

Four-leaved rose: $r = a \cos 2\theta$

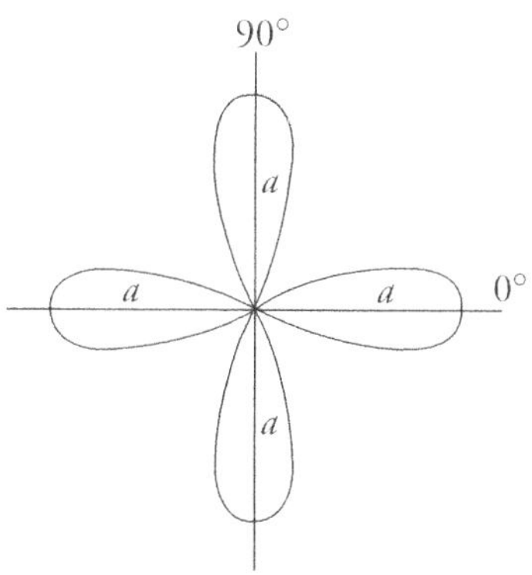

(n)

Four-leaved rose: $r = a \sin 2\theta$

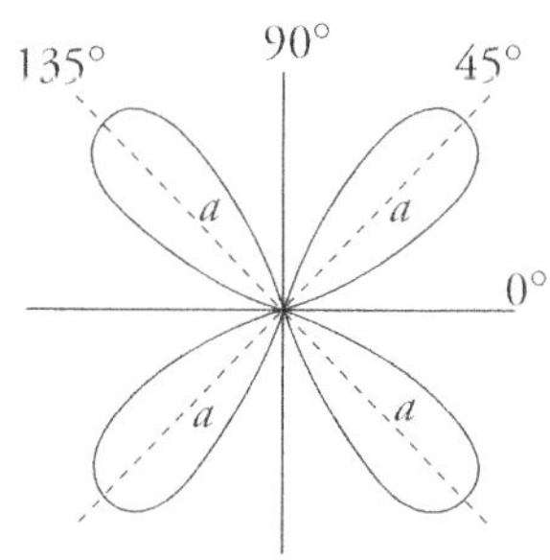

(o)

Complex Numbers

We know that squares are non-negative for real numbers (*Theorem 7.26*). Therefore, a square root of a negative real number is not real. However, if we consider square roots of negative real numbers as numbers, then we can make further progress in mathematics and related fields. We define $\sqrt{-1}$ as the number i so that $i^2 = -1$. We know that the number i is not a real number; therefore, it does not lie on the real number line. We introduce another number line perpendicular to the real number line at 0. We call this number line the **imaginary number line.** We place the number i at the point 1 unit from 0. Then the square roots of all negative numbers lie on this imaginary number line.

The plane with the rectangular coordinate system defined by the real number line (**real axis**) and the imaginary number line (imaginary axis) is called the **complex plane.**

If we look at real numbers and imaginary numbers as the endpoints of vectors in standard position, then by using vector addition, the sum of a real number and an imaginary number gives us a new vector, as shown below.

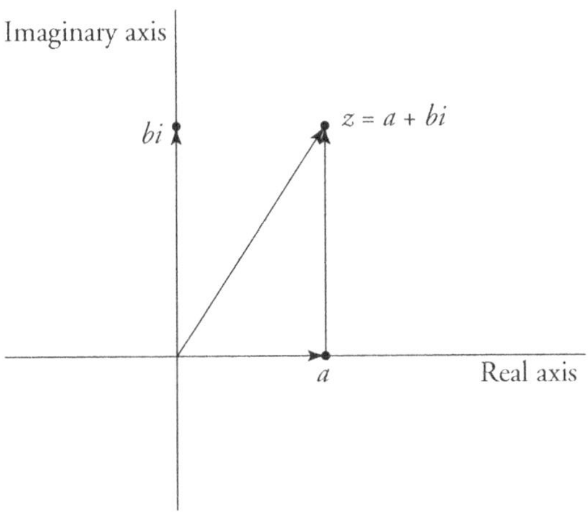

The endpoint of the new vector is a point on the complex plane that we identify as $a + bi$. Such a number is called a **complex number.** It is conventional to use the letter z to represent complex numbers. The real number a is called the *real part* of the complex number z and the real number b is called the *imaginary part* of the complex number z.

As you can see, the set of complex numbers is a larger set than the set of real numbers, as all real numbers reside on the real axis.

We define the sum of two complex numbers using vector addition:

If $z_1 = a_1 + b_1 i$ and $z_2 = a_2 + b_2 i$, then $z_1 + z_2 = (a_1 + a_2) + (b_1 + b_2)i$.

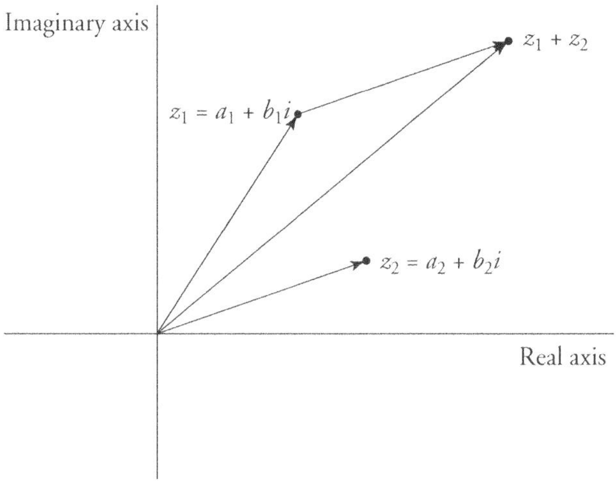

Treating i as a variable, complex numbers as linear polynomials of i, and using $i^2 = -1$, we define the multiplication of complex numbers as follows:

If $z_1 = a_1 + b_1 i$ and $z_2 = a_2 + b_2 i$, then $z_1 z_2 = (a_1 a_2 - b_1 b_2) + (a_1 b_2 + a_2 b_1)i$.

We assume all the properties of real numbers are true for complex numbers. That is, complex numbers satisfy the commutative property of addition, the commutative property of multiplication, the associative property of addition, the associative property of multiplication, and the distributive property.

We define a **complex conjugate** of a complex number as $z = a + bi$, denoted by $\bar{z}$, as $\bar{z} = a - bi$.

Example 10: Find the complex conjugate of $z = -3 - 2i$.

By the definition of the complex conjugate of a complex number, $\bar{z} = -3 + 2i$.

Theorem 9.2: If $z = a + bi$, then $z\bar{z} = a^2 + b^2$. In particular, $z\bar{z}$ is a real number.

Proof: By the definition of the multiplication of complex numbers,

$$\begin{aligned} z\bar{z} &= \left(a^2 + b^2\right) + \left(ab - ab\right)i \\ &= a^2 + b^2 \end{aligned}$$

Theorem 9.3: If $z_1 = a_1 + b_1 i$ and $z_2 = a_2 + b_2 i$, then

$$\frac{z_1}{z_2} = \left(\frac{a_1 a_2 + b_1 b_2}{a_2^{\,2} + b_2^{\,2}}\right) + \left(\frac{-a_1 b_2 + b_1 a_2}{a_2^{\,2} + b_2^{\,2}}\right) i$$

Proof:

$$\frac{z_1}{z_2} = \frac{a_1 + b_1 i}{a_2 + b_2 i}$$

Multiply both the numerator and the denominator by $\bar{z}_2$.

$$\begin{aligned} \frac{z_1}{z_2} &= \frac{\left(a_1 + b_1 i\right)\left(a_2 - b_2 i\right)}{\left(a_2 + b_2 i\right)\left(a_2 - b_2 i\right)} \\ &= \frac{\left(a_1 a_2 + b_1 b_2\right) + \left(-a_1 b_2 + a_2 b_1\right) i}{a_2^{\,2} + b_2^{\,2}} \\ &= \left(\frac{a_1 a_2 + b_1 b_2}{a_2^{\,2} + b_2^{\,2}}\right) + \left(\frac{-a_1 b_2 + b_1 a_2}{a_2^{\,2} + b_2^{\,2}}\right) i \end{aligned}$$

The next-to-last equality follows from the definition of multiplication and *Theorem 9.2*.

Polar Form of a Complex Number

Much more interesting formulas for multiplication and division of complex numbers can be obtained if we introduce a polar coordinate system to the complex plane.

Let the origin of the complex plane be the pole and the positive real axis be the polar axis. Let (r, θ) be the polar coordinates of the complex number $a + bi$ in this coordinate system, where $r \geq 0$ and $0 \leq \theta < 2\pi$.

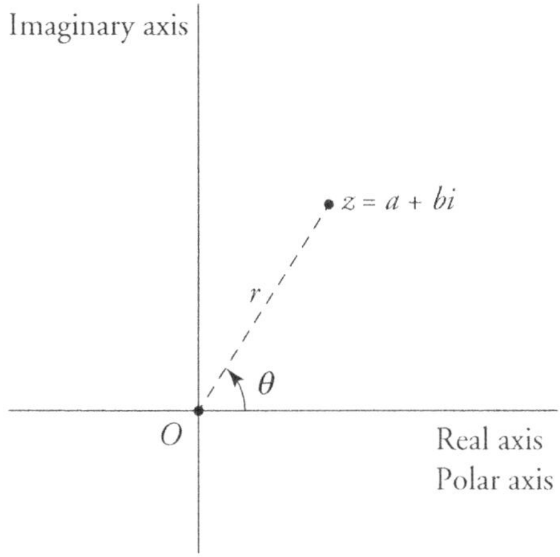

By Table 9-1, $a = r\cos\theta$, $b = r\sin\theta$, $r = \sqrt{a^2 + b^2}$, and $\tan\theta = \frac{b}{a}$. Therefore, z can be written as:

$$\begin{aligned} z &= a + bi \\ &= r\cos\theta + (r\sin\theta)i \\ &= r(\cos\theta + (\sin\theta)i) \end{aligned}$$

We can remove the extra parentheses if we write $(\sin\theta)\,i$ as $i\sin\theta$.

$z = r(\cos\theta + i\sin\theta)$, where $r \geq 0$ and $0 \leq \theta < 2\pi$, is called the **polar form** of the complex number z. The number r is called the **modulus** of z and the number θ is called the **argument** of z.

Example 11: Write the complex number $z = 3 - 3i$ in polar form.

The given complex number z is in the fourth quadrant of the complex plane. $r = \sqrt{3^2 + (-3)^2} = 3\sqrt{2}$ and $\tan\theta = -1$.

The general solution of the above equation is $\theta = n\pi + \tan^{-1}(-1)$, where n is an integer. That is, $\theta = n\pi - \frac{\pi}{4}$.

When $n = 2$, we get the solution in the fourth quadrant that satisfies the requirement $0 \leq \theta < 2\pi$. That is, $\theta = 2\pi - \frac{\pi}{4} = \frac{7\pi}{4}$. Therefore, $z = 3\sqrt{2}\left(\cos\frac{7\pi}{4} + i\sin\frac{7\pi}{4}\right)$.

Example 12: Write the complex number $z = -\sqrt{3} + i$ in polar form.

The given complex number z is in the second quadrant of the complex plane.

$$r = \sqrt{\left(\sqrt{3}\right)^2 + 1^2} = 2 \text{ and } \tan\theta = -\frac{1}{\sqrt{3}}$$

The general solution of the above equation is $\theta = n\pi + \tan^{-1}\left(-\frac{1}{\sqrt{3}}\right)$ where n is an integer. That is, $\theta = n\pi - \frac{\pi}{6}$.

When $n = 1$, we get the solution in the second quadrant that satisfies the requirement $0 \leq \theta < 2\pi$. That is, $\theta = \pi - \frac{\pi}{6} = \frac{5\pi}{6}$. Therefore, $z = 2\left(\cos\frac{5\pi}{6} + i\sin\frac{5\pi}{6}\right)$.

Multiplication and Division of Complex Numbers in Polar Form

Theorem 9.4: If $z = r(\cos\theta + i\sin\theta)$, then $z\bar{z} = r^2$.

Proof: By *Theorem 9.2*,

$$\begin{aligned} z\bar{z} &= a^2 + b^2 \\ &= (r\cos\theta)^2 + (r\sin\theta)^2 \\ &= r^2\left(\cos^2\theta + \sin^2\theta\right) \\ &= r^2 \end{aligned}$$

The last equality follows from *Theorem 2.8.*

Theorem 9.5: Let $z_1 = r_1\left(\cos\theta_1 + i\sin\theta_1\right)$ and $z_2 = r_2\left(\cos\theta_2 + i\sin\theta_2\right)$ be two complex numbers in polar form. Then, $z_1z_2 = r_1r_2\left(\cos\left(\theta_1 + \theta_2\right) + i\sin\left(\theta_1 + \theta_2\right)\right)$.

That is, the modulus of z_1z_2 is r_1r_2, and the argument is $\theta_1 + \theta_2$.

Proof:

$$z_1z_2 = \left[r_1\left(\cos\theta_1 + i\sin\theta_1\right)\right]\left[r_2\left(\cos\theta_2 + i\sin\theta_2\right)\right]$$
$$= r_1r_2\left(\cos\theta_1 + i\sin\theta_1\right)\left(\cos\theta_2 + i\sin\theta_2\right)$$
$$= r_1r_2\left(\left(\cos\theta_1\cos\theta_2 - \sin\theta_1\sin\theta_2\right) + i\left(\sin\theta_1\cos\theta_2 + \cos\theta_1\sin\theta_2\right)\right)$$

By *Theorems 7.7* (cosine sum) and *7.8* (cosine difference),
$z_1z_2 = r_1r_2\left(\cos\left(\theta_1 + \theta_2\right) + i\sin\left(\theta_1 + \theta_2\right)\right)$.

Theorem 9.6: Let $z_1 = r_1\left(\cos\theta_1 + i\sin\theta_1\right)$ and $z_2 = r_2\left(\cos\theta_2 + i\sin\theta_2\right)$ be two complex numbers in polar form. Then,
$\frac{z_1}{z_2} = \frac{r_1}{r_2}\left(\cos\left(\theta_1 - \theta_2\right) + i\sin\left(\theta_1 - \theta_2\right)\right)$.
That is, the modulus of $\frac{z_1}{z_2}$ is $\frac{r_1}{r_2}$, and the argument is $\theta_1 - \theta_2$.

Proof:

$$\frac{z_1}{z_2} = \frac{r_1\left(\cos\theta_1 + i\sin\theta_1\right)}{r_2\left(\cos\theta_2 + i\sin\theta_2\right)}$$

Multiply both the numerator and the denominator by $\overline{z}_2$.

$$\frac{z_1}{z_2} = \frac{r_1r_2\left(\cos\theta_1 + i\sin\theta_1\right)\left(\cos\theta_2 - i\sin\theta_2\right)}{r_2^{\,2}}$$

$$= \frac{r_1}{r_2}\left(\left(\cos\theta_1\cos\theta_2 + \sin\theta_1\sin\theta_2\right) + i\left(\sin\theta_1\cos\theta_2 - \cos\theta_1\sin\theta_2\right)\right)$$

$$\frac{z_1}{z_2} = \frac{r_1}{r_2}\left(\cos\left(\theta_1 - \theta_2\right) + i\sin\left(\theta_1 - \theta_2\right)\right)$$

The last equality follows from *Theorems 7.6* and *7.9*. The second equality follows from *Theorem 9.4*.

Example 13: If $z = 4(\cos 65° + i \sin 65°)$ and $w = 7(\cos 105° + i \sin 105°)$, find zw and $\frac{z}{w}$.

$$zw = (4)(7)\left[\cos(65°+105°)+i\sin(65°+105°)\right]$$

$$zw = 28(\cos 170°+i\sin 170°)$$

$$\frac{z}{w} = \frac{4}{7}\left[\cos(65°-105°)+i\sin(65°-105°)\right]$$

$$\frac{z}{w} = \frac{4}{7}\left[\cos(-40°)+i\sin(-40°)\right]$$

$$\frac{z}{w} = \frac{4}{7}\left[\cos(320°)+i\sin(320°)\right] \text{ in polar form}$$

De Moivre's Theorem and Powers of Complex Numbers in Polar Form

*Theorem 9.7 (**De Moivre's Theorem**):* If $z = r(\cos\theta + i\sin\theta)$, then $z^n = r^n(\cos n\theta + i\sin n\theta)$ for any positive integer n.

There are infinitely many statements that arise from this theorem, for each positive integer n.

For example, when $n = 7$, the statement of the theorem is $z^7 = r^7(\cos 7\theta + i\sin 7\theta)$. We use the following method, also known as "mathematical induction," to prove De Moivre's Theorem.

Proof: Suppose that the statement is true for some arbitrary positive integer k. That is, suppose $z^k = r^k(\cos k\theta + i\sin k\theta)$ is true.

Multiply both sides by z. Then we get:

$$z^k z = \left[r^k(\cos k\theta + i\sin k\theta)\right]\left[r(\cos\theta + i\sin\theta)\right]$$

$$z^{k+1} = \left[r^k(\cos k\theta + i\sin k\theta)\right]\left[r(\cos\theta + i\sin\theta)\right]$$

Then by *Theorem 9.5*,

$$z^{k+1} = r^{k+1}\left(\cos(k\theta+\theta)+i\sin(k\theta+\theta)\right)$$

$$= r^{k+1}\left(\cos(k+1)\theta+i\sin(k+1)\theta\right)$$

That is, if the statement of the theorem is true for some integer k, then it is also true for $k + 1$. Let us call this statement the inductive step.

Since $z = r(\cos\theta + i\sin\theta)$, the statement of the theorem is true for $k = 1$. Therefore, by the inductive step, the statement of the theorem is true for $k = 2$.

Since the statement of the theorem is true for $k = 2$, by the inductive step, the statement of the theorem is also true for $k = 3$, and so on, ad infinitum.

With De Moivre's Theorem in place, we can find integer powers of complex numbers.

Example 14: Find the polar form of $\left(\sqrt{3}+i\right)^7$.

First, convert $z = \sqrt{3}+i$ to polar form.

The number z is in the first quadrant.

$$r = \sqrt{\left(\sqrt{3}\right)^2 + 1^2} = 2 \text{ and } \tan\theta = \frac{1}{\sqrt{3}}$$

The solution of the basic tangent equation in the first quadrant is $\tan^{-1}\left(\frac{1}{\sqrt{3}}\right) = \frac{\pi}{6}$. Therefore, $z = 2\left(\cos\frac{\pi}{6} + i\sin\frac{\pi}{6}\right)$.

Now, by De Moivre's Theorem,

$$\begin{aligned} z^7 &= 2^7\left(\cos\frac{7\pi}{6} + i\sin\frac{7\pi}{6}\right) \\ &= 128\left(\cos\frac{7\pi}{6} + i\sin\frac{7\pi}{6}\right) \end{aligned}$$

Example 15: Write $\left(\sqrt{2}-i\sqrt{2}\right)^4$ in polar form.

Let $z = \sqrt{2}-i\sqrt{2}$. This number is in the fourth quadrant.

$$r = \sqrt{\left(\sqrt{2}\right)^2 + \left(-\sqrt{2}\right)^2} = 2$$

$$\tan\theta = -1$$

The general solution of the above equation is $\theta = n\pi + \tan^{-1}(-1)$, where n is an integer. That is, $\theta = n\pi - \frac{\pi}{4}$.

When $n = 2$, we get the solution $\theta = \frac{7\pi}{4}$, which is in the fourth quadrant and also satisfies the requirement $0 \le \theta < 2\pi$. Therefore, $z = 2\left(\cos\left(\frac{7\pi}{4}\right) + i\sin\left(\frac{7\pi}{4}\right)\right)$.

By De Moivre's Theorem, $z^4 = 2^4(\cos 7\pi + i \sin 7\pi)$.

However, 7π is not in the interval $[0, 2\pi]$. By using *Theorem 2.9*, we can see that $\cos 7\pi = \cos(6\pi + \pi) = \cos \pi$ and $\sin 7\pi = \sin(6\pi + \pi) = \sin \pi$. Therefore, $z^4 = 16(\cos \pi + i \sin \pi)$.

Complex Roots of Complex Numbers in Polar Form

Recall the definition of the **nth root** of a real number for a positive integer n. If a and b are real numbers and $a^n = b$, then we say a is the nth root of b. For example, since $2^3 = 8$, the third root of 8 is 2.

We define the nth root of a complex number the same way we defined the nth root of a real number. If z and w are complex numbers and $w^n = z$, where n is a positive integer, then we say w is a complex nth root of z.

Theorem 9.8: Let $z = r(\cos\theta + i\sin\theta)$. Then, $w_k = r^{\frac{1}{n}}\left(\cos\left(\frac{2k\pi+\theta}{n}\right) + i\sin n\left(\frac{2k\pi+\theta}{n}\right)\right)$ is an nth root of z, for any positive integers n and k.

Proof: By De Moivre's Theorem,

$$w_k{}^n = \left(r^{\frac{1}{n}}\right)^n\left(\cos n\left(\frac{2k\pi+\theta}{n}\right) + i\sin\left(\frac{2k\pi+\theta}{n}\right)\right)$$

$$= r\left(\cos(2k\pi+\theta) + i\sin(2k\pi+\theta)\right)$$

Now by *Theorem 2.9*, $\cos(2k\pi + \theta) = \cos\theta$ and $\sin(2k\pi + \theta) = \sin\theta$. Therefore,

$$\begin{aligned} w_k{}^n &= r\left(\cos\left(2k\pi+\theta\right)+i\sin\left(2k\pi+\theta\right)\right) \\ &= r\left(\cos\theta+i\sin\theta\right) \\ &= z \end{aligned}$$

Then, by definition, w is an nth root of z.

Some of the nth roots of z listed in *Theorem 9.8* may not be in polar form. It is possible that $\frac{2k\pi+\theta}{n} \geq 2\pi$, for some k. If we insist that the nth root of a complex number must be in polar form, then we have to put some restrictions on k. The following theorem lists *all* complex nth roots in polar form.

Theorem 9.9: Let $z = r(\cos\theta + i\sin\theta)$. Then, $w_k = r^{\frac{1}{n}}\left(\cos\left(\frac{2k\pi+\theta}{n}\right)+i\sin\left(\frac{2k\pi+\theta}{n}\right)\right)$ is an nth root of z in polar form for any positive integer n, and for any positive integer k, so that $0 \leq k \leq n-1$.

Proof: By *Theorem 9.8*, $w_k = r^{\frac{1}{n}}\left(\cos\left(\frac{2k\pi+\theta}{n}\right)+i\sin\left(\frac{2k\pi+\theta}{n}\right)\right)$ is an nth root of z, for any positive integer k.

However, for each w_k to be in polar form, $\frac{2k\pi+\theta}{n} < 2\pi$. This leads to

$$\begin{aligned} 2k\pi+\theta &< 2\pi n \\ 2k\pi &< 2\pi n-\theta \\ k &< n-\frac{\theta}{2\pi} \end{aligned}$$

But $\frac{\theta}{2\pi} < 1$, because $\theta < 2\pi$. Since both k and n are positive integers, this implies that $k \leq n-1$. Therefore, $0 \leq k \leq n-1$.

Example 16: Find the complex fourth roots of $z = \sqrt{3} + i$ in polar form.

We will convert z to polar form first.

The number z is in the first quadrant of the complex plane.

$$r = \sqrt{3+1} = 2 \text{ and } \tan\theta = \frac{1}{\sqrt{3}}$$

The solution of the tangent equation in the first quadrant is $\frac{\pi}{6}$. Therefore, $z = 2\left(\cos\left(\frac{\pi}{6}\right) + i\sin\left(\frac{\pi}{6}\right)\right)$.

Then by *Theorem 9.9*, the complex fourth roots of z are $w_k = 2^{\frac{1}{4}}\left(\cos\left(\frac{2k\pi + \frac{\pi}{6}}{4}\right) + i\sin\left(\frac{2k\pi + \frac{\pi}{6}}{4}\right)\right)$, where $k = 0, 1, 2,$ and 3.

That is, the complex fourth roots of z are as follows: $w_0 = 2^{\frac{1}{4}}\left(\cos\left(\frac{\pi}{24}\right) + i\sin\left(\frac{\pi}{24}\right)\right)$,

$$w_1 = 2^{\frac{1}{4}}\left(\cos\left(\frac{2\pi + \frac{\pi}{6}}{4}\right) + i\sin\left(\frac{2\pi + \frac{\pi}{6}}{4}\right)\right) = 2^{\frac{1}{4}}\left(\cos\left(\frac{13\pi}{24}\right) + i\sin\left(\frac{13\pi}{24}\right)\right),$$

$$w_2 = 2^{\frac{1}{4}}\left(\cos\left(\frac{4\pi + \frac{\pi}{6}}{4}\right) + i\sin\left(\frac{4\pi + \frac{\pi}{6}}{4}\right)\right) = 2^{\frac{1}{4}}\left(\cos\left(\frac{25\pi}{24}\right) + i\sin\left(\frac{25\pi}{24}\right)\right),$$

and $w_3 = 2^{\frac{1}{4}}\left(\cos\left(\frac{6\pi + \frac{\pi}{6}}{4}\right) + i\sin\left(\frac{6\pi + \frac{\pi}{6}}{4}\right)\right) = 2^{\frac{1}{4}}\left(\cos\left(\frac{37\pi}{24}\right) + i\sin\left(\frac{37\pi}{24}\right)\right).$

Example 17: Find the complex third roots of 1 in polar form.

Let $z = 1$. Then the polar form of z is $z = \cos 0 + i \sin 0$.

Then by *Theorem 9.9*, the complex third roots of z are $w_k = \left(\cos\left(\frac{2k\pi}{3}\right) + i\sin\left(\frac{2k\pi}{3}\right)\right)$, for $k = 0$, 1, and 2.

That is, the complex third roots of 1 are as follows: $w_0 = (\cos 0 + i \sin 0)$, $w_1 = \left(\cos\left(\frac{2\pi}{3}\right) + i\sin\left(\frac{2\pi}{3}\right)\right)$, and $w_2 = \left(\cos\left(\frac{4\pi}{3}\right) + i\sin\left(\frac{4\pi}{3}\right)\right)$.

The following figure shows the complex third roots of 1 on the complex plane.

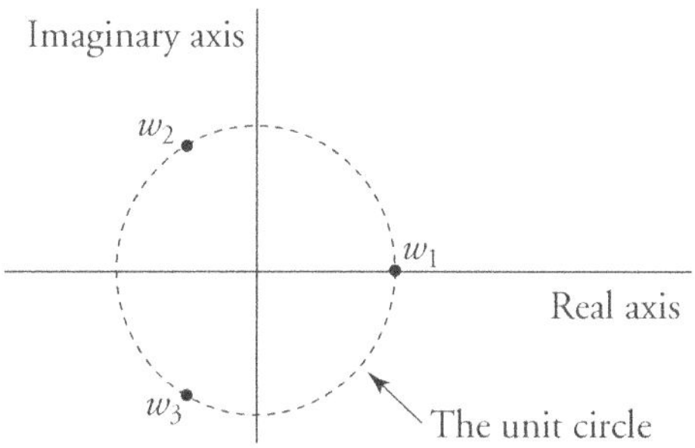

Chapter Check-Out

Questions

1. Convert $P(2, 5)$ from rectangular coordinates to polar coordinates.
2. True or False: The graph of $r = \sin \theta$ is a circle with center O.
3. Convert the complex number $-2 + 2i$ to polar coordinates.
4. If $\mathbf{z} = 2(\cos 70° + i \sin 70°)$ and $\mathbf{w} = 6(\cos 80° + i \sin 80°)$, find $\mathbf{zw}$.
5. Find the complex square roots of 1.

Answers

1. $\left(\sqrt{29}, 68.2°\right)$
2. False
3. $2\sqrt{2}\left(\cos\dfrac{3\pi}{4} + i\sin\dfrac{3\pi}{4}\right)$
4. **zw** $= 12(\cos 150° + i \sin 150°)$
5. $1, -1$

Chapter 10
ADDITIONAL TOPICS

Chapter Check-In

- ❑ Defining uniform circular motion
- ❑ Using uniform circular motion to solve problems about linear velocity
- ❑ Defining simple harmonic motion
- ❑ Solving problems using simple harmonic motion

Common Core Standard: Model periodic phenomena with trigonometric functions

Choose trigonometric functions to model periodic phenomena with specified amplitude, frequency, and midline. (HSF.TF.B.5)

Understand radian measure of an angle as the length of the arc on the unit circle subtended by the angle. (HSF.TF.A.1)

Use special triangles to determine geometrically the values of sine, cosine, and tangent for $\frac{\pi}{3}$, $\frac{\pi}{4}$, and $\frac{\pi}{6}$, and use the unit circle to express the values of sine, cosine, and tangent for x, $\pi + x$, and $2\pi - x$ in terms of their values for x, where x is any real number. (HSF.TF.A.3)

Please also refer to the standards for mathematical practice, found here: www.corestandards.org/Math/Practice/.

Uniform Circular Motion

If θ is the measure of a central angle of a circle measured in radians, then the length of the intercepted arc (s) can be found by multiplying the radius of the circle (r) by the measure of the central angle (θ): $s = r\theta$. Remember, θ must be measured in radians. (*See CliffsNotes Geometry Common Core Quick Review.*)

Example 1: Find the length (s) of the arc intercepted by a central angle of 3 radians if the radius of the circle is 5 centimeters.

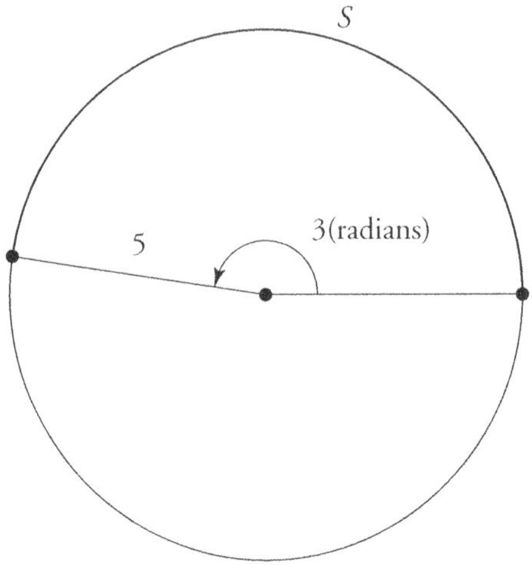

$$\begin{aligned} s &= r\theta \\ &= (5)(3) \\ &= 15 \end{aligned}$$

Thus, the length of the intercepted arc is 15 centimeters.

Suppose a point particle travels around a circle at a constant speed. Then we say that the particle is in **uniform circular motion.** If the particle traverses an arc of length s in t time units, then the **constant linear speed** (v) of the particle is the distance traveled by the particle in a given time period t. That is, $v = \frac{s}{t} = \frac{r\theta}{t}$, where θ is the central angle subtended by s.

Example 2: If the radius of Earth is 4,050 miles, what is the linear speed of an object located on the equator, given that Earth rotates one complete revolution (2π radians) in 24 hours? Round to the nearest mile.

$$\begin{aligned} v &= \frac{r\theta}{t} \\ &= \frac{(4{,}050)(2\pi)}{24} \\ &= \frac{2{,}025\pi}{6} \\ &\approx 1{,}060 \text{ mph} \end{aligned}$$

The above approximation is obtained using a scientific calculator.

Suppose a point particle travels around a circle at a constant speed. If it traverses an arc with a subtended central angle θ in t time units, then the **angular speed** (ω) of the point particle is $\omega = \frac{\theta}{t}$.

Example 3: Point P revolves counterclockwise around a circle with the center O, making seven complete revolutions in 5 seconds. If the radius of the circle is 8 centimeters, find the linear and angular speeds of point P. Approximate your answers to two decimal places.

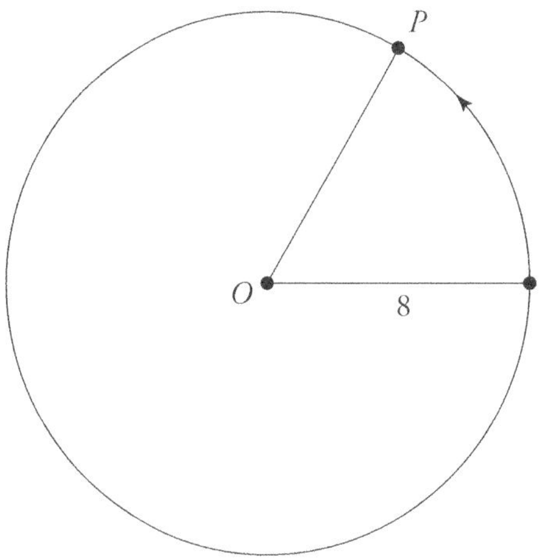

$$
\begin{aligned}
v &= \frac{s}{t} \\
&= \frac{7(2\pi)(8)}{5} \\
&= \frac{112\pi}{5} \\
&\approx 70.37
\end{aligned}
$$

Thus, the linear speed is approximately 70.37 centimeters per second.

$$\begin{aligned}\omega &= \frac{\theta}{t} \\ &= \frac{7(2\pi)}{5} \\ &= \frac{14\pi}{5} \\ &\approx 8.80\end{aligned}$$

Thus, the angular speed is approximately 8.80 radians per second.

Simple Harmonic Motion

Consider a point particle that is in uniform circular motion. The **velocity** of the particle at a given time is a vector whose magnitude is the linear speed at that time and whose direction is the unit vector tangent to the circle at that instant, as shown in the following figure.

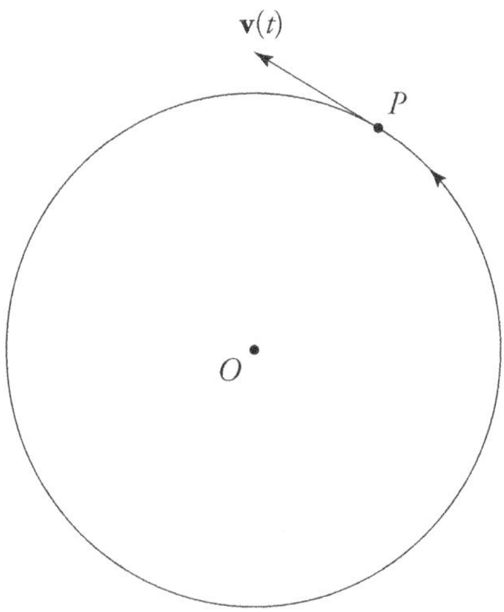

Let the velocity of the particle at an arbitrary time be $\mathbf{v}(t)$. We are using functional notation to say the velocity varies with time. The magnitude of the velocity is the linear speed. Since the linear speed is constant, the magnitude of the velocity does not change with time. Let the linear speed be v. Then $\|\mathbf{v}(t)\| = v$ for any time t.

Let $\mathbf{v}(t) = x(t)\mathbf{i} + y(t)\mathbf{j}$, where $x(t)\mathbf{i}$ is the component vector of $\mathbf{v}(t)$ in the x-direction and $y(t)\mathbf{j}$ is the component vector of $\mathbf{v}(t)$ in the y-direction.

Then by *Theorem 8.8*, $v = \sqrt{(x(t))^2 + (y(t))^2}$.

Let line $\overleftrightarrow{SR}$ be the horizontal line passing through O, as shown in the following figure. Even though v is constant, $x(t)$ and $y(t)$ are not constant with time. For example, at point R, $x(t) = 0$ and $y(t) = v$. As another example, at point S, $x(t) = 0$ and $y(t) = -v$.

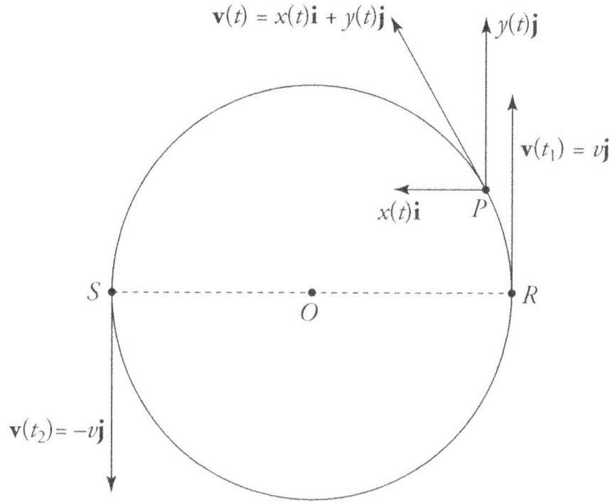

Now consider a particle Q that moves along the horizontal line $\overleftrightarrow{SR}$ with the speed $x(t)$ in concert with the particle P that is in uniform circular motion. See the figure below.

The motion of particle Q is limited to segment $\overline{SR}$.

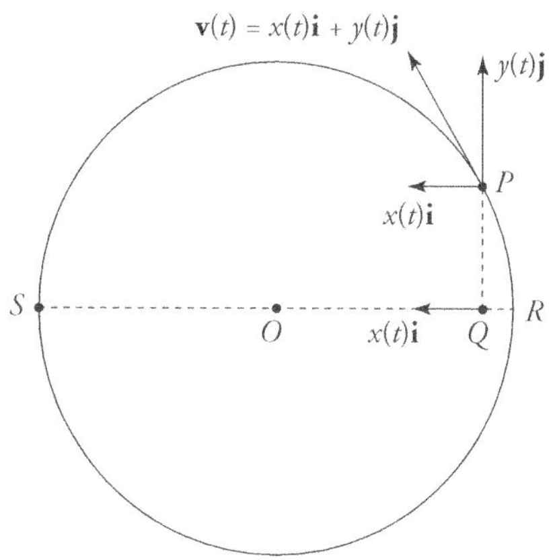

Suppose particle Q is initially at point R. That is, Q is at R when $t = 0$. The speed of Q when it is at point R is 0.

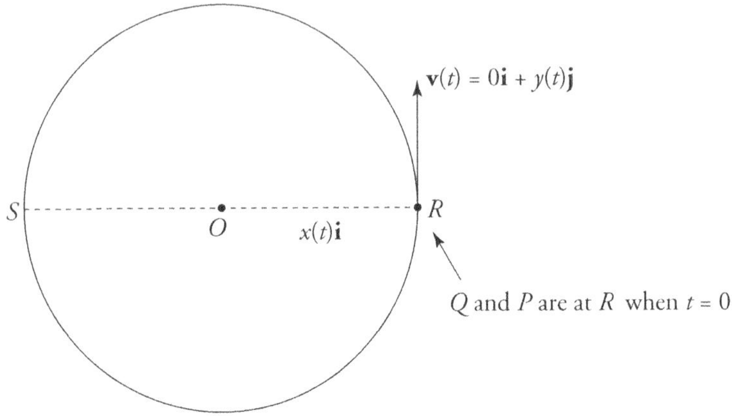

When moving from point R to O, the speed of particle Q increases; it then decreases when moving from O to S, and the speed of Q is 0 at S. The direction of the velocity of particle Q remains $-\mathbf{i}$ for the motion from R to S. As particle P traces the lower semicircle, particle Q turns and travels from S back to R.

From point S to O, the speed of particle Q increases; it then decreases from O to R, and the speed of Q is 0 again at R. The direction of the velocity of particle Q remains $\mathbf{i}$ for the motion from S to R.

The magnitude of $x(t)$ is linearly related to the length PQ (the y-coordinate of P). The length PQ is 0 at R, and then increases and reaches the maximum value (equal to the radius of the circle) when Q is at O. Then PQ decreases from O to S and is equal to 0 at S. The length PQ (the $-y$-coordinate of P) then increases again until Q reaches O, and then it decreases back to 0 when Q reaches R.

Therefore, we can say that the motion of particle Q is periodic and sinusoidal. That is, $x(t) = A \sin Bt$ for some constants A and B.

When particle Q is at O, $x(t) = v$ and from the equation above, $x(t)$ is maximum when $\sin Bt$ is maximum (i.e., when $\sin Bt = 1$). Therefore, the maximum value of $x(t)$ from the equation is A. Therefore, $A = v$.

Suppose the angular speed of P is ω. Then period T of the motion of Q is $T = \frac{2\pi}{\omega}$. By *Theorem 3.3*, period T of $x(t)$ is $T = \frac{2\pi}{B}$. Therefore, $B = \omega$. Therefore, the motion of particle Q is $x(t) = v \sin \omega t$.

The graph of one cycle of the motion is given below.

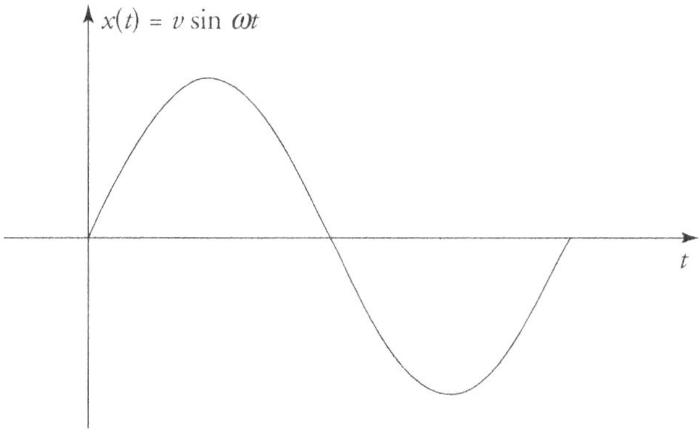

The **frequency** (f) of a motion is defined as $f = \frac{1}{T}$. That is, the frequency, by definition, is the reciprocal of the period and it gives us the number of cycles of the motion per unit time. In this case, the frequency is $f = \frac{\omega}{2\pi}$.

Suppose instead that particle Q is at point O when $t = 0$.

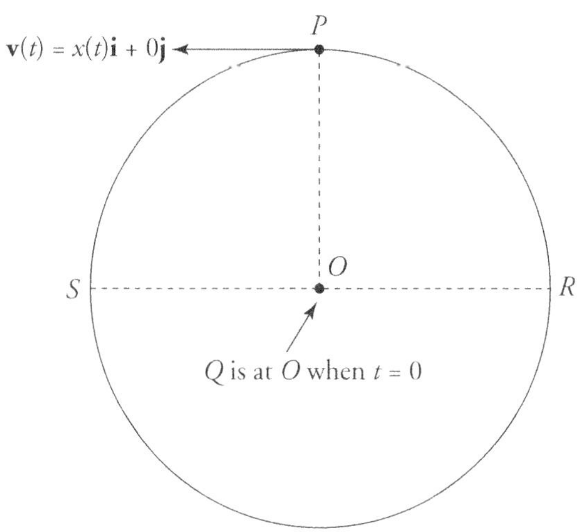

Then the initial speed of particle Q is v. When Q moves from O to S, the speed decreases from v to 0 and the direction of the motion is $-\mathbf{i}$. Then Q changes direction and travels to R. The speed increases back to v when it is at O and decreases to 0 when it reaches R. Then it turns and travels back to O, completing a full cycle. Now the motion of the particle is given by the equation $x(t) = v \cos \omega t$.

The graph of one cycle of the motion is given below.

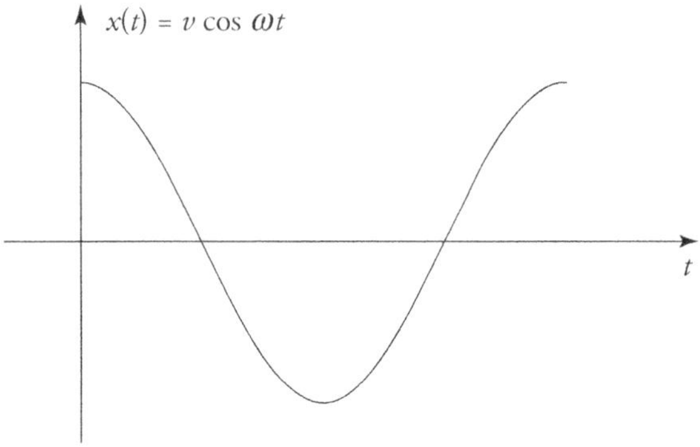

If the initial position is any point other than R, S, or O, then the initial speed is neither 0 nor v.

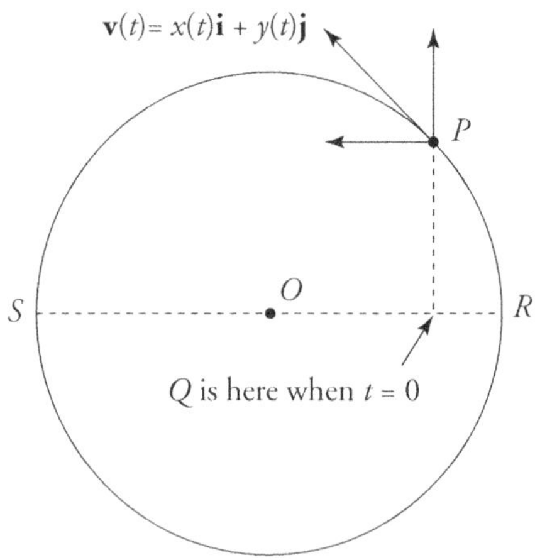

That means the speed of Q may first increase to the maximum value of v or it may decrease to 0. Therefore, the equation of the motion can have a **phase shift.**

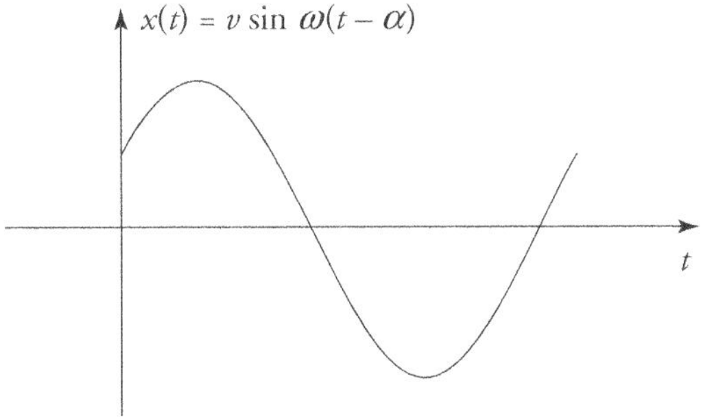

Motions that satisfy the equations $d(t) = A \cos B(t - C)$ or $d(t) = A \sin B(t - C)$, where d is the displacement at a given time t, A and B are constants determined by the specific motion, and t is a measurement of time, are referred to as *simple harmonic motions.*

Example 4: Suppose one end of a spring is attached to a wall and the other end is attached to a mass. Suppose the mass is pulled 2 centimeters from its rest position at $t = 0$, and then let go. Due to the restoration force in the spring, assume that the mass will move in a simple harmonic motion. If it takes 6 seconds for the mass to come back to the position it was in when $t = 0$, find the amplitude, period, frequency, and equation of the motion.

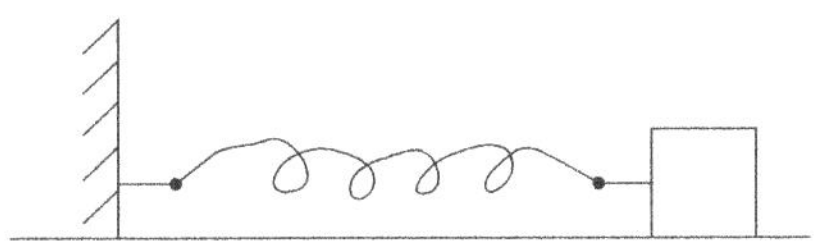

Let the displacement of the mass be $x(t)$, where the mass was at $x = 0$, before it was pulled.

The amplitude of the motion is 2 centimeters, the period is 6 seconds, and the frequency is $\frac{1}{6}$ cycles per second. Since $x(0) = 2$, the motion is represented by a cosine function. The equation of the motion is

$$x(t) = 2\cos\left(\frac{2\pi}{6}\right)t$$

$$x(t) = 2\cos\frac{1}{3}\pi t$$

Example 5: The horizontal displacement $d(t)$ of the end of a pendulum undergoes a simple harmonic motion given by

$$d(t) = K \sin(2\pi t)$$

Find the *amplitude* K of the motion if $d(3.25) = 12$. That is, $d = 12$ centimeters when $t = 3.25$ seconds.

$$\begin{aligned} d(t) &= K \sin(2\pi t) \\ 12 &= K \sin(2\pi(3.25)) \\ K &= \frac{12}{\sin 6.5\pi} \\ &= \frac{12}{\sin\left(6\pi + \frac{\pi}{2}\right)} \\ &= \frac{12}{\sin\frac{\pi}{2}} \\ &= 12 \end{aligned}$$

The next-to-last equality comes from *Theorem 2.9.*

Chapter Check-Out

Questions

1. Find the length of an arc intercepted by a central angle of size 2.3 radians if the circle has a radius of 12 inches.
2. If a ball with a radius of 2 feet is spinning at 12 rpm, what is the linear velocity of a point on the equator of the ball?
3. If a point revolves around a circle with a radius of 12 at a constant rate of 4 revolutions every 2 minutes, find its angular velocity.
4. If the displacement of a spring (d) is given by $d = A \cos \alpha t$ where A, the initial displacement, and d are expressed in inches, t in seconds, and $\alpha = 8$, find d when A is 10 inches and $t = 2\pi$.

Answers

1. 27.6 inches
2. $\dfrac{48\pi \text{ feet}}{\text{min}}$
3. $\omega = \dfrac{\theta}{t} = \dfrac{4(2\pi)}{2} = 4\pi$
4. 10 inches

REVIEW QUESTIONS

Use this review to practice what you've learned in this book and to build your confidence in working with trigonometry. After you work through the review questions, you'll be well on your way to achieving your goal of being proficient in trigonometry.

Questions

Chapter 1

1. True or False: A dilation preserves the lengths of segments.
2. State the converse of the Pythagorean Theorem.
3. True or False: If the sides of a triangle have lengths 11, 60, and 61, then the triangle is a right triangle.

Chapter 2

4. Find the exact value of cos270°. Do not use a calculator.
5. Find the exact value of (cos 60°). Do not use a calculator.
6. Determine the sign of the following trigonometric functions:

   a) sin 255°

   b) tan 240°

   c) cos (–110°)
7. Find sin θ and cos θ for the acute angle θ if $\tan\theta = \dfrac{1}{2\sqrt{2}}$.
8. Find sin θ and cos θ if θ is in the third quadrant and tan $\theta = 1$.
9. For a circle with radius 3 feet, find the arc length *s* subtended by a central angle of 6°.
10. Convert 171° to radian measure. Give the exact answer.

11. What is the exact value of $\sin\frac{\pi}{4} - \tan\frac{\pi}{4}$? Do not use a calculator.

A. $\frac{2\sqrt{3}-3}{6}$

B. $\frac{1}{2}$

C. $\frac{-\sqrt{3}}{2}$

D. $\frac{\sqrt{2}-2}{2}$

12. Find the exact value of $\tan\left(-\frac{2}{3}\pi\right)$.

Chapter 3

13. Find the amplitude, the period, and the phase shift of $f(x) = 3\sin 2\theta$.

14. Find the amplitude, the period, and the phase shift of

$f(x) = 3\cos 2\left(\theta + \frac{\pi}{3}\right)$.

15. Find the amplitude, the period, and the phase shift of $f(x) = \tan 2\theta$.

Chapter 4

16. What is the exact value of $\tan\left(\cos^{-1}\left(\frac{1}{2}\right)\right)$?

A. $\sqrt{3}$

B. $\frac{\sqrt{3}}{2}$

C. $\frac{1}{2}$

D. $\frac{\sqrt{3}}{3}$

17. Find the exact value of $\cos(\cos^{-1} \circ 1)$.

18. What is the exact value of $\sin^{-1}\left(\sin\frac{5\pi}{4}\right)$?

19. What is $\sin(\sin^{-1} \circ 1)$?

Chapter 5

20. Find the general solution of $\cos\theta = \frac{1}{2}$.

21. Find the general solution of $\sin\theta = \frac{1}{2}$.

22. Find the general solution of $\tan\theta = 1$.

23. Find particular solutions of $\cos\theta = \frac{1}{2}$ in the interval $[0, 2\pi]$.

24. Find particular solutions of $\sin\theta = \frac{1}{2}$ in the interval $[0, 2\pi]$.

25. Find particular solutions of $\tan\theta = 1$ in the interval $[0, 2\pi]$.

Chapter 6

26. Solve this triangle: $\alpha = 51°$, $\beta = 49°$, $b = 70$.

27. Solve this triangle: $a = 10$, $b = 11$, $\alpha = 27°$.

28. Solve this triangle: $b = 7$, $c = 4$, $\beta = 94°$.

29. What is the solution to this triangle: $a = 11$, $b = 17$, $c = 14$?

A. $\alpha = 84.78°$, $\beta = 40.12°$, $\gamma = 55.10°$
B. $\alpha = 55.10°$, $\beta = 84.78°$, $\gamma = 40.12°$
C. $\alpha = 40.12°$, $\beta = 84.78°$, $\gamma = 55.10°$
D. $\alpha = 40.12°$, $\beta = 55.10°$, $\gamma = 84.78°$

30. What is the area of a triangle with $a = 68°$, $b = 7$ feet, and $c = 7$ feet?

A. 9.18 square feet
B. 45.43 square feet
C. 22.72 square feet
D. 24.50 square feet

31. Find the area of a triangle with sides 4 meters, 5 meters, and 6 meters.

Chapter 7

32. Is the following statement an identity? $\dfrac{\cos x}{1-\sin x} = \sec x + \tan x$

33. If $\sin A = \frac{3}{7}, \frac{\pi}{2} \le A \le \pi$ and $\cos B = -\frac{5}{8}, \pi \le B \le \frac{3\pi}{2}$, find the exact value of $\cos(A + B)$.

34. Establish this identity: $\cos\left(\theta + \frac{\pi}{2}\right) = -\sin\theta$.

35. Find the exact value of $\sin 2\theta$ if $\sin\theta = \frac{3}{5}, \frac{\pi}{2} < \theta < \pi$.

36. The expression $\csc 2\theta + \cot 2\theta$ forms an identity with which of the following?

A. $\tan\theta$

B. $\tan\frac{\theta}{2}$

C. $\cot\theta$

D. $\cot\frac{\theta}{2}$

37. Which of the following correctly expresses $\sin 2\theta\cos 7\theta$ as a sum containing only sines or cosines?

A. $\frac{1}{2}(\sin 9\theta + \sin 5\theta)$

B. $\frac{1}{2}(\sin 9\theta - \sin 5\theta)$

C. $\sin\frac{9\theta}{2} - \sin\frac{5\theta}{2}$

D. $\sin 5\theta + \sin 2\theta$

Chapter 8

38. What is the position vector of the vector $v = \overrightarrow{JK}$ if $J = (5,-4)$ and $K = (-6,-7)$?

A. $11\mathbf{i} + 3\mathbf{j}$

B. $-11\mathbf{i} + 3\mathbf{j}$

C. $-11\mathbf{i} - 3\mathbf{j}$

D. $-2\mathbf{i} - 11\mathbf{j}$

39. If $\mathbf{v} = -3\mathbf{i} + 7\mathbf{j}$ and $\mathbf{w} = 5\mathbf{i} + 10\mathbf{j}$, what is the value of $\mathbf{v} - \mathbf{w}$?

A. $-8\mathbf{i} + -3\mathbf{j}$

B. $2\mathbf{i} - 17\mathbf{j}$

C. $-8\mathbf{i} + 3\mathbf{j}$

D. $2\mathbf{i} + 17\mathbf{j}$

40. If $\mathbf{v} = 3\mathbf{i} + 2\mathbf{j}$ and $\mathbf{w} = 3\mathbf{i} + 8\mathbf{j}$, find $5\mathbf{v} - 3\mathbf{w}$ and $||5\mathbf{v} - 3\mathbf{w}||$.

41. What is the unit vector with the same direction as $\mathbf{v} = -10\mathbf{i} + 24\mathbf{j}$?

A. $-\frac{5}{13}\mathbf{i} + \frac{12}{13}\mathbf{j}$

B. $-\frac{10}{\sqrt{34}}\mathbf{i} + \frac{24}{\sqrt{34}}\mathbf{j}$

C. $-\frac{5}{17}\mathbf{i} + \frac{12}{17}\mathbf{j}$

D. $-\mathbf{i} + \mathbf{j}$

42. If $\mathbf{v} = 5\mathbf{i} - 9\mathbf{j}$ and $\mathbf{w} = 27\mathbf{i} + 15\mathbf{j}$, what is $\mathbf{v} \cdot \mathbf{w}$?

A. 270

B. –450

C. 360

D. 0

43. What is the measure of the angle between the vectors $\mathbf{v} = \mathbf{i} + 4\mathbf{j}$ and $\mathbf{w} = -3\mathbf{i} + 4\mathbf{j}$?

A. 73°

B. 129°

C. 51°

D. 49°

44. Find the vector $\mathbf{v}$ given $||v|| = 6$ and the angle between the direction of $\mathbf{v}$ and the positive x-axis is $\alpha = 150°$.

45. Which of the following vectors is orthogonal to $-10\mathbf{i} + 6\mathbf{j}$?

A. $7\mathbf{i} + 4\mathbf{j}$

B. $2\mathbf{i} + 2\mathbf{j}$

C. $-12\mathbf{i} - 20\mathbf{j}$

D. $-15\mathbf{i} + 9\mathbf{j}$

Chapter 9

46. What are the rectangular coordinates of $(9, 30°)$?

A. $\left(\dfrac{9\sqrt{3}}{2}, \dfrac{9}{2}\right)$

B. $\left(\dfrac{9\sqrt{2}}{2}, \dfrac{9}{2}\right)$

C. $\left(\dfrac{9}{2}, \dfrac{9\sqrt{3}}{2}\right)$

D. $\left(-\dfrac{9\sqrt{3}}{2}, \dfrac{9}{2}\right)$

47. Find the polar coordinates of $(-12, -12)$ for $r > 0$, $0 \le \theta < 2\pi$.

48. Determine the polar form of the complex number $2 - 4i$. Express the angle θ in degrees where $0 \le \theta < 360°$, and round numerical entries to two decimal places.

Chapter 10

49. A propeller measuring 2 meters from tip to tip is rotating at a rate of 400 revolutions per minute. Find the linear velocity of a tip of the propeller in meters per minute.

50. The horizontal displacement of a pendulum is described by the equation $d = K\sin 2\pi t$. Find K if $d = 20$ and $t = \dfrac{1}{6}$.

51. The horizontal displacement of a pendulum is described by the equation $d = K\sin 2\pi t$. Find d if $t = 1.2$ and $K = 11$.

Answers

Chapter 1

1. False
2. If the sum of the squares of the lengths of two of the sides is equal to the square of the length of the third side, then the triangle is a right triangle.
3. True

Chapter 2

4. 0
5. $\frac{1}{4}$
6. a) negative; b) positive; c) negative
7. $\cos\theta = \frac{2\sqrt{2}}{3} \sin\theta = \frac{1}{3}$
8. $\sin\theta = -\frac{1}{\sqrt{2}}; \quad \cos\theta = -\frac{1}{\sqrt{2}}$
9. 0.31 feet
10. $\frac{19}{20}\pi$
11. D
12. $\sqrt{3}$

Chapter 3

13. amplitude = 3; period = π; phase shift = 0
14. amplitude = 3; period = π; phase shift = $-\frac{\pi}{3}$
15. no amplitude; period = $\frac{\pi}{2}$; phase shift = 0

Chapter 4

16. A

17. $^{\circ}1$

18. $\frac{\pi}{4}$

19. Doesn't exist

Chapter 5

20. $2n\pi \pm \frac{\pi}{3}$, where n is an integer

21. $n\pi + (-1)^n \frac{\pi}{6}$, where n is an integer

22. $n\pi + \frac{\pi}{4}$, where n is an integer

23. $\frac{\pi}{3}, \frac{5\pi}{3}$

24. $\frac{\pi}{6}, \frac{5\pi}{6}$

25. $\frac{\pi}{4}, \frac{5\pi}{4}$

Chapter 6

26. $\alpha = 51°, \beta = 49°, \gamma = 80°, a \approx 72.08, b = 70, c \approx 91.34$

27. There are two solutions:

(1) $\alpha = 27°, \beta \approx 29.96°, \gamma \approx 123.04°, a = 10, b = 11, c \approx 18.46$

(2) $\alpha = 27°, \beta \approx 150.04°, \gamma \approx 2.96°, a = 10, b = 11, c \approx 1.14$

28. $\alpha \approx 51.25°, \beta = 94°, \gamma \approx 34.75°, a \approx 5.47, b = 7, c = 4$

29. C

30. C

31. 9.92 square meters

Chapter 7

32. Yes

33. $\frac{1}{56}\left(10\sqrt{10}+3\sqrt{39}\right)$

34. $\cos\left(\theta+\frac{\pi}{2}\right)=\cos\theta\cos\frac{\pi}{2}-\sin\theta\sin\frac{\pi}{2}=(\cos\theta)(0)-\sin\theta(1)=-\sin\theta$

35. $-\frac{24}{25}$

36. C

37. B

Chapter 8

38. C

39. A

40. $5\mathbf{v}-3\mathbf{w}=6\mathbf{i}-14\mathbf{j}$; $\|5\mathbf{v}-3\mathbf{w}\|=2\sqrt{58}$

41. A

42. D

43. C

44. $-3\sqrt{3}\mathbf{i}+3\mathbf{j}$

45. C

Chapter 9

46. A

47. $\left(12\sqrt{2},\ \frac{5\pi}{4}\right)$

48. $4.47(\cos 296.57° + i\sin 296.57°)$

Chapter 10

49. 800π meters per minute

50. $\frac{40\sqrt{3}}{3}$

51. 10.46

GLOSSARY

AA Criteria for Similarity: Two triangles with two pairs of equal angles are similar.

amplitude of a periodic function: Half of the absolute value of the difference between the maximum value of the trigonometric function and the minimum value of the trigonometric function, if exists.

angle: A rotation around the origin from a ray identified as the initial side to a ray identified as the terminal side.

(angle in) standard position: When an angle is placed on an *xy*-coordinate system so that the vertex is at the origin and the initial side is on the positive *x*-axis.

angle of depression: An angle measured below the horizontal.

angle of elevation: An angle measured above the horizontal.

angular speed: Speed defined in terms of angle of rotation and time.

argument (of a complex number): When a complex number is written in polar form, the second polar coordinate, that is θ, is called the argument of the complex number.

ASA Theorem: Given two triangles ΔABC and ΔDEF so that $\angle A = \angle D$, $\angle B = \angle E$, and $AB = DE$, the triangles are congruent.

bearing: An angle measured clockwise from due north to a vector.

complex conjugate (of a complex number): If a complex number z is $a + bi$, then the complex conjugate $\overline{z}$ is $a - bi$, where a and b are real numbers.

complex number: A number of the form $a + bi$, where a and b are real numbers and $i^2 = 1$, is called a complex number.

complex plane: A coordinate system for complex numbers.

congruence: A congruence in the plane is a plane transformation that is equal to the composition of a finite number of basic rigid motions.

constant linear speed: If d is the distance traveled by a particle in a given period of time t, then the constant linear speed is the ratio $\frac{d}{t}$.

converse of the Pythagorean Theorem: Suppose ΔABC is a right triangle so that $a^2 + b^2 = c^2$. Then ΔABC is a right triangle with $\angle C = 90°$.

coterminal: Two angles in standard position that share a terminal side.

degree measure of an angle: Place a circle on the angle so that the center of the circle coincides with the vertex of the angle. Partition the circumference into 360 equal parts. Each part is called a degree. The number of degrees in the arc of the circle subtended by the angle is called the degree measure of the angle.

De Moivre's Theorem: If $z = r(\cos\theta + i\sin\theta)$, then $z^n = r^n(\cos n\theta + i\sin n\theta)$ for any positive integer n.

diagonal: The graph of the line $y = x$.

dilation: A dilation is a plane transformation with a fixed point O (known as the center) and a positive scale factor r so that any point P on the plane other than O will get transformed to a point P' so that $OP' = r \cdot OP$.

direction of a vector: The direction of a non-zero vector **u** is $\frac{1}{\|\mathbf{u}\|}\mathbf{u}$. The direction of the zero vector is arbitrary.

dot product: An operation combining two vectors to yield a single number.

equal vectors: Two vectors that have the same magnitude and direction.

even function: A function is even if $f(-x) = f(x)$.

frequency: The frequency of a simple harmonic is the reciprocal of the period of the motion. It is the number of cycles of the motion per unit time.

general solution: Solutions of an equation over all real numbers.

Heron's formula: If s is the semi perimeter of a triangle, then the area of the triangle is $\sqrt{s(s-a)(s-b)(s-c)}$, where a, b, and c are the lengths of the three sides of the triangle.

HL Theorem: If two right triangles have equal hypotenuses and one pair of equal legs, then the two triangles are congruent.

imaginary axis or **imaginary number line:** An axis in the complex plane.

initial point (of a vector): This is one of the two endpoints of a vector when the vector is represented by a line segment. Together with the other endpoint (known as the terminal point), the two points are used to indicate the direction of the vector.

initial side: Side of an angle where angle measurement begins.

Law of Cosines: A relationship between the lengths of the three sides of a triangle and the cosine of one of the angles.

Law of Sines: A relationship between the ratios of the sines of the angles of a triangle and the sides opposite those angles.

magnitude of a vector: One of the two properties of a vector. When a vector is identified by a line segment with endpoints identified as the initial point and the terminal point, then the length of the segment represents the magnitude of the vector.

maximum value: Largest value of a function.

midline of a periodic function: A line that runs between the maximum and minimum values of the function.

minimum value: Smallest value of a function.

modulus (of a complex number): Square root of the sum of the squares of the complex number's real and imaginary coefficients.

negative angle: Results from clockwise rotation.

***n*th root:** If z and w are complex numbers so that $z = w^n$ for some positive integer n, then w is called an nth root of z.

odd function: A function is odd if $f(-x) = -f(x)$.

one-to-one (1-1): A characteristic of some functions where each element in the domain is paired with one and only one element in the range, and vice versa.

orthogonal vectors: Two vectors are orthogonal if the angle between them is $\frac{\pi}{2}$.

period: The period of a periodic function is the smallest number p so that $f(x + p) = f(x)$, for all real numbers x. By definition, this is the length of the interval $[x, x + p]$.

periodic function: A function f is periodic if there is a number p so that $f(x + p) = f(x)$ for all x.

phase shift: The horizontal displacement of a function to the right or left of the vertical axis.

polar axis: A ray extending from the pole in a polar coordinate system.

polar coordinates: An ordered pair consisting of a radius and an angle.

polar coordinate system: A coordinate system using distance and angle to identify points.

polar form (of a complex number): When a complex number z is written in the form $z = r(\cos\theta + i \sin\theta)$, where $r \geq 0$ and $0 \leq \theta < 2\pi$.

pole: The fixed center of the polar coordinate system.

position vector: Another name for a standard vector.

positive angle: Results from counterclockwise rotation.

Pythagorean Theorem: Let ΔABC be a right triangle with $\angle C = 90°$. Then $a^2 + b^2 = c^2$.

quadrantal angle: An angle in standard position with its terminal side on a coordinate axis.

radian measure of an angle: Place a circle on the angle so that the center of the circle coincides with the vertex of the angle. The ratio of the arc length of the arc subtended by the angle and the radius of the circle is called the radian measure.

real axis or **real number line:** An axis in the complex plane.

SAA Theorem: Given two triangles ΔABC and ΔDEF so that $AC = DF$, $\angle A = \angle D$, and $\angle B = \angle E$, the triangles are congruent.

SAS Theorem: Given two triangles ΔABC and ΔDEF so that $\angle A = \angle D$, $AB = DE$, and $AC = DF$, the triangles are congruent.

scalar: A scalar is a real number.

scalar multiplication: An operation that changes the magnitude of a vector.

semi-perimeter: One-half the perimeter of a triangle.

similar: Two figures are similar if there is a similarity that carries one figure on top of the other.

similarity: The composition of a congruence F followed by a dilation D.

special angles: Angles with degree measure 0°, 30°, 45°, 60°, and 90°.

SSS Theorem: Two triangles with three equal sides that are congruent.

SSS Theorem for Similarity: If ΔABC and ΔPQR are triangles so that

$$\frac{AB}{PQ} = \frac{BC}{QR} = \frac{CA}{RP},$$

then $\Delta ABC \sim \Delta PQR$.

static equilibrium: An object is said to be in a state of static equilibrium if all the forces acting on the object add up to zero.

terminal point: The endpoint of a vector.

terminal side: The side of an angle where the angle measurement ends.

uniform circular motion: A particle is in uniform circular motion if its linear speed is a constant for any given time interval.

unit circle: The circle with center (0, 0) and radius 1.

unit vector: A unit vector is a vector with a magnitude of 1.

vector: A vector is a mathematical concept with just two properties, namely the magnitude and the direction.

vector addition: The process of combining two vectors.

velocity: A velocity of a moving particle is a vector whose magnitude is the speed of the particle.

vertical asymptote: A vertical line at which the function is undefined and increases or decreases without bounds just to the left of the line or just to the right of the line.

zero vector: A vector with a magnitude of zero and any direction.

Appendix
THEOREMS

There are numerous trigonometry theorems given in this book. They are numbered for organizational purposes. As a handy study reference, the theorems are compiled in this appendix.

Chapter 1: Geometric Prerequisites

Theorem 1.1: A congruence:
(i) maps lines to lines, rays to rays, and segments to segments.
(ii) maps line segments to line segments of equal length.
(iii) maps an angle to an angle of the same degree.

Theorem 1.2: A line segment is congruent to a line segment of equal length.

Theorem 1.3: An angle is congruent to an angle of the same degree.

Theorem 1.4 (SAS Theorem): Given two triangles ΔABC and ΔDEF so that $\angle A = \angle D$, $AB = DE$, and $AC = DF$, the triangles are congruent.

Theorem 1.5 (ASA Theorem): Given two triangles ΔABC and ΔDEF so that $\angle A = \angle D$, $\angle B = \angle E$, and $AB = DE$, the triangles are congruent.

Theorem 1.6 (SSS Theorem): Two triangles with three equal sides are congruent.

Theorem 1.7 (HL Theorem): If two right triangles have equal hypotenuses and one pair of equal legs, then the two triangles are congruent.

Theorem 1.8: A dilation:
(i) maps lines to lines, rays to rays, and segments to segments.
(ii) maps a line passing through the center O to itself.
(iii) maps a line not passing through O to a line parallel to it.

Theorem 1.9: A dilation preserves degrees of angles.

Theorem 1.10: Any two circles are similar.

Theorem 1.11 (AA Criteria for Similarity): Two triangles with two pairs of equal angles are similar.

Theorem 1.12: If two triangles are similar, then the corresponding angles are congruent.

Theorem 1.13 (SSS Theorem for Similarity): If ΔABC and ΔPQR are triangles so that $\frac{AB}{PQ} = \frac{BC}{QR} = \frac{CA}{RP}$, then $\Delta ABC \sim \Delta PQR$.

Theorem 1.14: Suppose ΔABC and ΔPQR are triangles so that $\Delta ABC \sim \Delta PQR$. Then $\frac{AB}{PQ} = \frac{BC}{QR} = \frac{CA}{RP}$.

Theorem 1.15 (SAS Criteria for Similarity): Given two triangles ΔABC and ΔPQR, if $\angle A = \angle P$ and $\frac{AB}{PQ} = \frac{AC}{PR}$, then $\Delta ABC \sim \Delta PQR$.

Theorem 1.16 (Pythagorean Theorem): Let ΔABC be a right triangle with $\angle B = 90°$. Then $AC^2 = AB^2 + BC^2$.

Theorem 1.17 (Converse of the Pythagorean Theorem): Suppose ΔABC is a right triangle so that $AB^2 + BC^2 = AC^2$. Then ΔABC is a right triangle with $\angle B = 90°$.

Chapter 2: Trigonometric Numbers of Angles

Theorem 2.1: Let θ be an angle in standard position. Then θ and $\theta + 2n\pi$ are coterminal angles for any integer n.

Theorem 2.2: Let θ be an angle in standard position with a positive measure less than $\frac{\pi}{2}$. Then both sin θ and cos θ are unique.

Theorem 2.3: Suppose θ is an acute angle of a right triangle. Then

$$\sin\theta = \frac{\text{length of the opposite side of } \theta}{\text{length of the hypotenuse}} \quad \text{and}$$

$$\cos\theta = \frac{\text{length of the adjacent side of } \theta}{\text{length of the hypotenuse}}.$$

Theorem 2.4: For any angle θ in standard position, the numbers sin θ and cos θ, as defined above, are unique.

Theorem 2.5: Suppose θ is an angle in standard position. Then the following are true:
(1) The angle θ is in the first quadrant if and only if $\sin\theta > 0$ and $\cos\theta > 0$.
(2) The angle θ is in the second quadrant if and only if $\sin\theta > 0$ and $\cos\theta < 0$.
(3) The angle θ is in the third quadrant if and only if $\sin\theta < 0$ and $\cos\theta < 0$.
(4) The angle θ is in the fourth quadrant if and only if $\sin\theta < 0$ and $\cos\theta > 0$.

Theorem 2.6: Any point P on the unit circle has coordinates $(\cos\theta, \sin\theta)$ for some angle θ in standard position, where P is the point of intersection of the unit circle and the terminal side of θ.

Theorem 2.7: If α and β are coterminal angles, then $\cos\alpha = \cos\beta$ and $\sin\alpha = \sin\beta$.

Theorem 2.8: For any angle θ, $\sin^2\theta + \cos^2\theta = 1$.

Theorem 2.9: For any angle θ and for any integer n:
(1) $\sin(2n\pi + \theta) = \sin\theta$ and
(2) $\cos(2n\pi + \theta) = \cos\theta$.

Theorem 2.10: For any angle θ:
(1) $\sin(-\theta) = -\sin\theta$ and
(2) $\cos(-\theta) = \cos\theta$.

Theorem 2.11: For any angle θ:
(1) $\sin(\pi - \theta) = \sin\theta$ and
(2) $\cos(\pi - \theta) = -\cos\theta$.

Theorem 2.12: $\sin 0 = 0$ and $\cos 0 = 1$.

Theorem 2.13: $\sin\frac{\pi}{2} = 1$ and $\cos\frac{\pi}{2} = 0$.

Theorem 2.14: $\sin\frac{\pi}{4} = \frac{\sqrt{2}}{2}$ and $\cos\frac{\pi}{4} = \frac{\sqrt{2}}{2}$.

Theorem 2.15: $\sin\frac{\pi}{6} = \frac{1}{2}$ and $\cos\frac{\pi}{6} = \frac{\sqrt{3}}{2}$.

Theorem 2.16: $\sin\frac{\pi}{3}=\frac{\sqrt{3}}{2}$ and $\cos\frac{\pi}{3}=\frac{1}{2}$.

Theorem 2.17: Consider a right triangle with an acute angle θ. Then $\tan\theta=\frac{\text{length of the opposite side of }\theta}{\text{length of the adjacent side of }\theta}$.

Theorem 2.18: Suppose θ is an angle in standard position and $P(x, y)$ is a point on the terminal side of θ other than O. Then, $\tan\theta=\frac{y}{x}$, $\csc\theta=\frac{r}{y}$, $\sec\theta=\frac{r}{x}$, and $\cot\theta=\frac{x}{y}$, if exists, where $r = OP$.

Theorem 2.19: For a given angle θ, $\cot\theta=\frac{\cos\theta}{\sin\theta}$ if $\sin\theta \neq 0$.

Chapter 3: Trigonometric Functions

Theorem 3.1: For any angle θ, $-1 \leq \sin\theta \leq 1$.

Theorem 3.2: For any angle θ, $-1 \leq \cos\theta \leq 1$.

Theorem 3.3: Suppose $f(x) = A\sin B(x - C)$, where A and B are non-zero constants and C is a constant. Then the amplitude of f is $|A|$, the period is $\frac{2\pi}{|B|}$, and the phase shift is $+C$ if $C > 0$ or $-|C|$ if $C < 0$.

Theorem 3.4: Suppose $f(x) = A\cos B(x - C)$, where A and B are non-zero constants and C is a constant. Then the amplitude of f is $|A|$, the period is $\frac{2\pi}{|B|}$, and the phase shift is $+C$ if $C > 0$ or $-|C|$ if $C < 0$.

Theorem 3.5: Suppose $f(x) = A\sin B(x - C) + D$, where A and B are non-zero constants and C and D are constants. Then the amplitude of f is $|A|$, the period is $\frac{2\pi}{|B|}$, the phase shift is $+C$ if $C > 0$ or $-|C|$ if $C < 0$, and the midline is $y = D$, where $y = f(x)$.

Theorem 3.6: Suppose $f(x) = A\cos B(x - C) + D$, where A and B are non-zero constants and C and D are constants. Then the amplitude of f is $|A|$, the period is $\frac{2\pi}{|B|}$, the phase shift is $+C$ if $C > 0$ or $-|C|$ if $C < 0$, and the midline is $y = D$, where $y = f(x)$.

Chapter 4: Inverse Trigonometric Functions

Theorem 4.1: Let $P(a, b)$ be a point on the xy-coordinate plane. Then the reflection of P across the line $y = x$ is the point (b, a).

Theorem 4.2: Suppose f is a 1-1 function. Then the graph of the inverse function is the reflection of the graph of f across the diagonal.

Chapter 5: Basic Trigonometric Equations

Theorem 5.1: The general solution of $\sin\theta = a$ is $n\pi + (-1)^n \sin^{-1} a$, where n is an integer and $-1 \le a \le 1$.

Theorem 5.2: The general solution of $\cos\theta = a$ is $2n\pi \pm \cos^{-1} a$, where n is an integer and $-1 \le a \le 1$.

Theorem 5.3: The general solution of the equation $\tan\theta = a$ is $n\pi + \tan^{-1}(a)$, where n is an integer.

Chapter 6: Trigonometry of Triangles

Theorem 6.1 (Pythagorean Theorem): If ΔABC is a right triangle so that $\angle C$ is the right angle, then $a^2 + b^2 = c^2$.

Theorem 6.2 (Converse of the Pythagorean Theorem): If ΔABC is a triangle so that $a^2 + b^2 = c^2$, then ΔABC is a right triangle and $\angle C$ is the right angle.

Theorem 6.3: In a triangle, a side facing a larger angle is longer than a side facing a smaller angle.

Theorem 6.4: In a triangle, an angle facing a longer side is larger than an angle facing a shorter side.

Theorem 6.5 (Triangle Inequality Theorem): In a triangle, the sum of any two sides is longer than the third side.

Theorem 6.6 (SAS Theorem for Congruence): Given two triangles ΔABC and ΔDEF so that $\angle A = \angle D$, $AB = DE$, and $AC = DF$, the triangles are congruent.

Theorem 6.7 (SAA Theorem for Congruence): Given two triangles ΔABC and ΔDEF so that $AC = DF$, $\angle A = \angle D$, and $\angle B = \angle E$, the triangles are congruent.

Theorem 6.8 (SSS Theorem for Congruence): Two triangles with three equal sides are congruent.

Theorem 6.9 (HL Theorem for Congruence): If two right triangles have equal hypotenuses and one pair of equal legs, then the two triangles are congruent.

Theorem 6.10: The sum of the degrees of angles of a triangle is 180°.

Theorem 6.11: Let θ be an acute angle of a right triangle and suppose the length of the hypotenuse is r. Then the length of the adjacent side of θ is $r \cos \theta$ and the length of the opposite side of θ is $r \sin \theta$.

Theorem 6.12 (Law of Cosines): Let ΔABC be a triangle with the adopted convention. Then,

$$a^2 = b^2 + c^2 - 2bc \cos \alpha$$

$$b^2 = a^2 + c^2 - 2ac \cos \beta$$

$$c^2 = a^2 + b^2 - 2ab \cos \gamma$$

Theorem 6.13 (Law of Sines): Let ΔABC be a triangle with the adopted convention. Then, $\dfrac{a}{\sin \alpha} = \dfrac{b}{\sin \beta} = \dfrac{c}{\sin \gamma}$.

Theorem 6.14: Let ΔABC be a triangle with the adopted convention. Let h_1 be the length of the perpendicular from point A to line $\overleftrightarrow{BC}$, let h_2 be the length of the perpendicular from point B to line $\overleftrightarrow{AC}$, and let h_3 be the length of the perpendicular from point C to line $\overleftrightarrow{AB}$. Then, the area of the triangle $= \dfrac{1}{2}ah_1 = \dfrac{1}{2}bh_2 = \dfrac{1}{2}ch_3$.

Theorem 6.15: Let ΔABC be a triangle with the accepted convention. Then, the area of the triangle $= \dfrac{1}{2}ab \sin \gamma = \dfrac{1}{2}bc \sin \alpha = \dfrac{1}{2}ca \sin \beta$.

Theorem 6.16 (Heron's Formula Theorem): If a, b, and c are the lengths of three sides of a triangle, and s is the semi-perimeter, then the area of the triangle is $\sqrt{s(s-a)(s-b)(s-c)}$.

Chapter 7: Trigonometric Identities and Equations

Theorem 7.1: $1 + \tan^2 \theta = \sec^2 \theta$.

Theorem 7.2: $\sin^2 \theta = 1 - \cos^2 \theta$.

Theorem 7.3: $\cos^2 \theta = 1 - \sin^2 \theta$.

Theorem 7.4: $\sin\left(\dfrac{\pi}{2} + \theta\right) = \cos\theta$ and $\cos\left(\dfrac{\pi}{2} + \theta\right) = -\sin\theta$.

Theorem 7.5: In a circle, if the central angles subtended by two arcs are equal in measure, then their corresponding chords are equal in measure.

Theorem 7.6: $\cos(\alpha - \beta) = \cos\alpha\cos\beta + \sin\alpha\sin\beta$.

Theorem 7.7: $\cos(\alpha + \beta) = \cos\alpha\cos\beta - \sin\alpha\sin\beta$.

Theorem 7.8: $\sin(\alpha + \beta) = \sin\alpha\cos\beta + \cos\alpha\sin\beta$.

Theorem 7.9: $\sin(\alpha - \beta) = \sin\alpha\cos\beta - \cos\alpha\sin\beta$.

Theorem 7.10: $\tan(\alpha+\beta) = \dfrac{\tan\alpha + \tan\beta}{1 - \tan\alpha\tan\beta}$.

Theorem 7.11: $\tan(\alpha-\beta) = \dfrac{\tan\alpha - \tan\beta}{1 + \tan\alpha\tan\beta}$.

Theorem 7.12: $\sin 2\theta = 2\sin\theta\cos\theta$.

Theorem 7.13: $\cos 2\theta = \cos^2\theta - \sin^2\theta$.

Theorem 7.14: $\cos 2\theta = 2\cos^2\theta - 1$.

Theorem 7.15: $\cos 2\theta = 1 - 2\sin^2\theta$.

Theorem 7.16: $\cos^2\theta = \dfrac{1+\cos 2\theta}{2}$ and $\sin^2\theta = \dfrac{1-\cos 2\theta}{2}$.

Theorem 7.17: $\tan 2\theta = \dfrac{2\tan\theta}{1-\tan^2\theta}$.

Theorem 7.18: $\sin\left(\dfrac{\theta}{2}\right) = \pm\sqrt{\dfrac{1-\cos\theta}{2}}$. The sign of the identity depends on the quadrant in which $\dfrac{\theta}{2}$ is located.

Theorem 7.19: $\cos\left(\dfrac{\theta}{2}\right) = \pm\sqrt{\dfrac{1+\cos\theta}{2}}$. The sign of the identity depends on the quadrant in which $\dfrac{\theta}{2}$ is located.

Theorem 7.20: $\tan\left(\dfrac{\theta}{2}\right) = \pm\sqrt{\dfrac{1-\cos\theta}{1+\cos\theta}}$. The sign of the identity depends on the quadrant in which $\dfrac{\theta}{2}$ is located.

Theorem 7.21: $\tan\frac{\theta}{2}=\frac{\sin\theta}{1+\cos\theta}$.

Theorem 7.22: $\tan\frac{\theta}{2}=\frac{1-\cos\theta}{\sin\theta}$.

Theorem 7.23 (Product-to-Sum Identities Theorem):

$$(1)\ \cos\alpha\cos\beta=\frac{1}{2}[\cos(\alpha+\beta)+\cos(\alpha-\beta)]$$

$$(2)\ \sin\alpha\sin\beta=-\frac{1}{2}[\cos(\alpha+\beta)-\cos(\alpha-\beta)]$$

$$(3)\ \sin\alpha\cos\beta=\frac{1}{2}[\sin(\alpha+\beta)+\sin(\alpha-\beta)]$$

$$(4)\ \cos\alpha\sin\beta=\frac{1}{2}[\sin(\alpha+\beta)-\sin(\alpha-\beta)]$$

Theorem 7.24 (Sum-to-Product Identities Theorem):

$$(1)\cos x+\cos y=2\cos\frac{x+y}{2}\cos\frac{x-y}{2}$$

$$(2)\cos x-\cos y=-2\sin\frac{x+y}{2}\sin\frac{x-y}{2}$$

$$(3)\sin x+\sin y=2\sin\frac{x+y}{2}\cos\frac{x-y}{2}$$

$$(4)\sin x-\sin y=2\cos\frac{x+y}{2}\sin\frac{x-y}{2}$$

Theorem 7.25 (Zero-Product Property Theorem): Suppose a and b are two real numbers. If $ab = 0$, then either $a = 0$ or $b = 0$.

Theorem 7.26 (Squares Are Non-Negative Theorem): If x is a real number, then $x^2 \geq 0$.

Theorem 7.27 (Square Root Principle Theorem): Suppose $x^2 = a$ for some nonnegative real number a. Then $x=\pm\sqrt{a}$.

Theorem 7.28 (Quadratic Formula Theorem): The solutions of the equation $ax^2 + bx + c = 0$, where a, b, and c are real constants, are $x=\frac{-b\pm\sqrt{b^2-4ac}}{2a}$, if $b^2 - 4ac \geq 0$.

Chapter 8: Vectors

Theorem 8.1: If **u** and **v** are two vectors, then **u** + **v** = **v** + **u**.

Theorem 8.2: If m and n are scalars and **u** is a vector, then $m(n\mathbf{u}) = n(m\mathbf{u}) = (mn)\mathbf{u}$.

Theorem 8.3: Let **u** be a non-zero vector. Then $\frac{1}{\|\mathbf{u}\|}\mathbf{u}$ is a unit vector.

Theorem 8.4: Suppose **u** is a non-zero vector. Then $\|\mathbf{u}\|\left(\frac{1}{\|\mathbf{u}\|}\mathbf{u}\right) = \mathbf{u}$.

Theorem 8.5: Let **u** and **v** be two non-zero vectors. Then the vector **u** can be written as a sum of two vectors orthogonal to each other so that one of the two vectors has the same direction as **v** or –**v**.

Theorem 8.6: Suppose $\mathbf{u} = \langle u_1, u_2\rangle$ and $\mathbf{v} = \langle v_1, v_2\rangle$ are two vectors. Then **u** = **v** if and only if $u_1 = v_1$ and $u_2 = v_2$.

Theorem 8.7: Let $\mathbf{u} = \langle u_1, u_2\rangle$ be a vector in standard position. Then $\mathbf{u} = u_1\mathbf{i} + u_2\mathbf{j}$. The vector $u_1\mathbf{i}$ is the vector component of **u** in the direction of **i** and $u_2\mathbf{j}$ is the vector component of **u** in the direction of **j**.

Theorem 8.8: If $\mathbf{u} = \langle u_1, u_2\rangle$, then $\|\mathbf{u}\| = \sqrt{u_1^{\,2} + u_2^{\,2}}$.

Theorem 8.9: Let $\mathbf{u} = \langle u_1, u_2\rangle$ and $\mathbf{v} = \langle v_1, v_2\rangle$. Then, $\mathbf{u} + \mathbf{v} = \langle u_1 + v_1, u_2 + v_2\rangle$.

Theorem 8.10: Let k be a scalar and $\mathbf{u} = \langle u_1, u_2\rangle$ be vector. Then the vector $k\mathbf{u} = \langle ku_1, ku_2\rangle$.

Theorem 8.11: Let $\mathbf{u} = \langle u_1, u_2\rangle$ and $\mathbf{v} = \langle v_1, v_2\rangle$. Then $\mathbf{u} - \mathbf{v} = \langle u_1 - v_1, u_2 - v_2\rangle$.

Theorem 8.12: If **u**, **v**, and **w** are vectors, then **u** + (**v** + **w**) = (**u** + **v**) + **w**.

Theorem 8.13: If **u** and **v** are vectors and k is a scalar, then $k(\mathbf{u} + \mathbf{v}) = k\mathbf{u} + k\mathbf{v}$.

Theorem 8.14: If m and n are scalars and **v** is a vector, then $(m + n)\mathbf{v} = m\mathbf{v} + n\mathbf{v}$.

Theorem 8.15: If **v** is a vector, then **v** + **0** = **0** + **v** = **v**.

Theorem 8.16: If **v** is a vector, then 1**v** = **v**.

Theorem 8.17: If **v** is a vector, then **v** + (–**v**) = **0**.

Theorem 8.18: If **u** and **v** are vectors, then **u** • **v** = **v** • **u**.

Theorem 8.19: If **u**, **v**, and **w** are vectors, then **u** • (**v** + **w**) = **u** • **v** + **u** • **w**.

Theorem 8.20: If **u** and **v** are vectors and k is a scalar, then **u** • (k**v**) = (k**u**) • **v** = k(**u** • **v**).

Theorem 8.21: If **u** is a vector, then **u** • **u** = $\|\mathbf{u}\|^2$.

Theorem 8.22: If **u** and **v** are vectors and θ is the (measure of the) angle between them, then **u** • **v** = $\|\mathbf{u}\|\ \|\mathbf{v}\|\cos\theta$.

Theorem 8.23: If **u** and **v** are two non-zero vectors, then
(1) The angle between **u** and **v** is acute if and only if **u** • **v** > 0.
(2) The angle between **u** and **v** is obtuse if and only if **u** • **v** < 0.
(3) **u** and **v** are orthogonal if and only if **u** • **v** = 0.

Theorem 8.24: Let **u** be a non-zero vector in standard position and let θ be the measure of the angle between **u** and **i**. Then **u** = $\langle \|\mathbf{u}\|\cos\theta, \|\mathbf{u}\|\sin\theta \rangle$.

Chapter 9: Polar Coordinates and Complex Numbers

Theorem 9.1: The equation of a circle with center (h, k) and radius $r > 0$ is $(x - h)^2 + (y - k)^2 = r^2$.

Theorem 9.2: If $z = a + bi$, then $z\bar{z} = a^2 + b^2$. In particular, $z\bar{z}$ is a real number.

Theorem 9.3: If $z_1 = a_1 + b_1 i$ and $z_2 = a_2 + b_2 i$, then

$$\frac{z_1}{z_2} = \left(\frac{a_1a_2 + b_1b_2}{a_2{}^2 + b_2{}^2}\right) + \left(\frac{-a_1b_2 + b_1a_2}{a_2{}^2 + b_2{}^2}\right)i.$$

Theorem 9.4: If $z = r(\cos\theta + i\sin\theta)$, then $z\bar{z} = r^2$.

Theorem 9.5: Let $z_1 = r_1(\cos\theta_1 + i\sin\theta_1)$ and $z_2 = r_2(\cos\theta_2 + i\sin\theta_2)$ be two complex numbers in polar form. Then,
$z_1z_2 = r_1r_2(\cos(\theta_1 + \theta_2) + i\sin(\theta_1 + \theta_2))$.

Theorem 9.6: Let $z_1 = r_1(\cos\theta_1 + i\sin\theta_1)$ and $z_2 = r_2(\cos\theta_2 + i\sin\theta_2)$ be two complex numbers in polar form. Then,

$$\frac{z_1}{z_2} = \frac{r_1}{r_2}(\cos(\theta_1 - \theta_2) + i\sin(\theta_1 - \theta_2)).$$

Theorem 9.7 (De Moivre's Theorem): If $z = r(\cos\theta + i\sin\theta)$, then $z^n = r^n(\cos n\theta + i\sin n\theta)$ for any positive integer n.

Theorem 9.8: Let $z = r(\cos\theta + i\sin\theta)$. Then,

$w_k = r^{\frac{1}{n}}\left(\cos\left(\frac{2k\pi+\theta}{n}\right) + i\sin n\left(\frac{2k\pi+\theta}{n}\right)\right)$ is an nth root of z, for any positive integers n and k.

Theorem 9.9: Let $z = r(\cos\theta + i\sin\theta)$. Then,

$w_k = r^{\frac{1}{n}}\left(\cos\left(\frac{2k\pi+\theta}{n}\right) + i\sin\left(\frac{2k\pi+\theta}{n}\right)\right)$ is an nth root of z in polar form for any positive integer n, and for any positive integer k, so that $0 \le k \le n - 1$.

INDEX

U

V

Z

Made in the USA
Monee, IL
13 September 2026

62981881R00154